THE
GLOBAL
AND
THE
INTIMATE

Feminism

in

Our

Time

Edited by

Geraldine **PRATT**

and

Victoria **ROSNER**

COLUMBIA UNIVERSITY PRESS   *NEW YORK*

Columbia University Press

Publishers Since 1893

New York Chichester, West Sussex

Copyright © 2012 Columbia University Press

All rights reserved

Library of Congress Cataloging-in-Publication Data

The global and the intimate : feminism in our time / edited by Geraldine
Pratt and Victoria Rosner.

   p.   cm.—(Gender and culture)

Includes bibliographical references.

ISBN 978-0-231-15448-2 (cloth : alk. paper)—ISBN 978-0-231-15449-9
(pbk. : alk. paper)—ISBN 978-0-231-52084-3 (e-book)

   1. Feminism—Case studies.   2. Intimacy (Psychology)—Case studies.

   I. Pratt, Geraldine.   II. Rosner, Victoria.

   HQ1154.G55   2012

   305.42—dc23

                          2012003963

c 10 9 8 7 6 5 4 3 2 1

p 10 9 8 7 6 5 4 3 2 1

References to Internet Web sites (URLs) were accurate at the
time of writing. Neither the author nor Columbia University
Press is responsible for Web sites that may have expired or
changed since the book was prepared.

The chapters "Jamaica Kincaid's Practical Politics of the
Intimate in *My Garden (book):*," by Agnese Fidecaro;
"Narratives and Rights: *Zlata's Diary* and the Circulation of
Stories of Suffering Ethnicity," by Sidonie Smith; and
"'Security Moms' in Twenty-First Century U.S.A.: The
Gender of Security in Neoliberalism," by Inderpal Grewal
have been revised from articles that originally appeared in
WSQ: Women's Studies Quarterly 34.1/2 (2006).

THE **GLOBAL** AND THE **INTIMATE**

Gender and Culture

CONTENTS

IV GLOBAL FEMINISM AND THE SUBJECTS OF KNOWLEDGE
||||||||||||||||||||||||||||||||||||||||

ACKNOWLEDGMENTS

Our friendship and collaboration began with an invitation to coedit an issue of *Women's Studies Quarterly*; we thank Cindi Katz and Nancy K. Miller for their intellectual matchmaking that brought us together. From tentative long-distance telephone calls between two strangers, in which we searched for shared sympathies and interests, the theme of the global and the intimate emerged. Our pleasure working together on the *WSQ* volume and our enthusiasm for the topic led to this second project, organized around the same theme; we are grateful to Gloria Jacobs for permission to reprint some essays here. We thank Jennifer Gieseking for arranging our first meeting at Mount Holyoke College in 2007, where we copresented introductory remarks at the Global and the Intimate conference and developed preliminary thoughts toward this volume and our introduction to it. We thank Rachel Adams for reading a first draft of the introduction and our anonymous referees for their helpful critical comments on the introduction and the manuscript as a whole. Thanks also to our series editor, Nancy K. Miller (once more!). Columbia University Press has provided a wonderful editorial home for this work; we are especially grateful to Jennifer Crewe. Our beautiful cover image appears courtesy of the artist, Mirta Kupferminc. Bonnie Kaserman and Sarah Zell helped us in preparing the manuscript, and the Canadian Social Sciences and Humanities Research Council provided financial support.

THE **GLOBAL** AND THE **INTIMATE**

INTRODUCTION

THE GLOBAL AND THE INTIMATE

Geraldine Pratt and Victoria Rosner

How many intimate relationships do you maintain through the Internet or over the phone? Can you touch those you love on a daily basis, or are you as likely to "hear" their voices through a text message or telephone call and see their faces on a computer screen? Do you live in the land of your birth, or did you or previous generations of your family originate in distant lands? Where did the food come from that you ingested today? When you rest your hands on your computer in the morning, do you ever think about where it was made and who made it? In a time of shrinking distances, virtual relationships, and increasing transnational interdependencies, the global and the intimate are more interwoven than ever. The people we care about, the things we put on or in our bodies, the money we earn, and the networks in which we circulate are ever more likely to collapse our habitual distinctions between near and far, strange and familiar, local and global.

Our pairing in this volume of the intimate and the global extends a long-standing feminist tradition of challenging gender-based oppositions by upending hierarchies of space and scale. Familiar feminist slogans like "the personal is the political" work by juxtaposing apparently incommensurate registers of experience and showing how categories defined in opposition to one another can produce pernicious exclusions. To disrupt traditional organizations of space, to forge productive dislocations, to reconfigure conventions of scale: these are the goals that underwrite many feminist investigations. In our own historical moment of growing global consciousness, they have led us to examine more carefully the ways in which the global and the intimate, typically imagined as mutually exclusive spheres, are profoundly intertwined.

Linking the global and the intimate does more than extend a common feminist practice; it may allow feminists to steer a somewhat different course through processes of globalization and relations of intimacy. Feminists have already noted the ways in which rhetorics of intimacy are built into the familiar local/global binary in the academic literature on globalization. This binary has been thoroughly criticized by critics such as Doreen Massey, Inderpal Grewal and Caren Kaplan, Sallie Marston, and J. K. Gibson-Graham[1] because it imaginatively constructs the local as a defense against powerful global forces in a way that seems to confirm the force and inevitability of certain modes of global capitalist expansion. By the same token, the local is often conceived as a defense against the forces of global capitalism in ways that call up established gender hierarchies of feminine and masculine. Local/global is a straightforward binary opposition, juxtaposing two terms that occupy polar positions on a single map. Intimate/global, we would like to suggest, both is and is not a binary; while irrefutably a pairing, it juxtaposes and resists the flattening effect of most binaries by bringing together terms that, we will argue, offer an implicit critique of one another.

In shifting from the idea of the local to that of the intimate, we are employing terms that are not defined *against* one another but rather draw their meaning from more elliptically related domains. We step out of hierarchical ontologies of scale or frameworks of micro/macro or the general and specific.[2] Intimacy is thus potentially and productively disruptive of the geographical binaries and hierarchies that often structure our thinking. By pairing the global and the intimate we aim to expose the patterns that recur when gender, sex, and the global imaginary combine. We have asked our contributors to show how the intimate and the global intertwine, to try and intervene in grand narratives of global relations by focusing on the specific, the quotidian, the affective, and the eccentric. We have also asked them to scrutinize the frames that we use to recognize and organize intimacy and to hunt for the global forces that quietly undergird personal experience and exchange.

Why is this exploration of the relationships between the global and the intimate a feminist project? Because we believe that the methodologies of feminism have proven particularly adept at exposing the universalist assumptions of international relations, social theory, and geographical models. Because feminism has a track record of success with slicing through the sometimes impersonal rhetorics of academic discourse. Because the sensitivities and suspicions that feminism has bred in us are part of what led us and our contributors to turn a skeptical eye on the occasionally uncritical affirmations of globalization. Because the same training has kept alive our collective belief that the analytical lens of gender relations can help to produce a more nuanced, complete, and just account of the world we share.

The form of this volume, no less than the topic, has feminist origins. Feminism, in the form of an invitation to collaborate from the editors of the journal *WSQ: Women's Studies Quarterly*, led us to this project. That invitation was predicated in part on an interest in bringing together two scholars from the disparate disciplines of geography and English, and it led to a special issue of *WSQ* where we first explored this volume's themes. That same cross-disciplinary feminist impetus prompted us to envision this volume as a blend of personal and critical writings, to collect essays from a diverse set of contributors, and to ask them to instantiate feminism as the crucial third term between the global and the intimate.

To invoke the global, we realize, risks reinscribing a set of pernicious assumptions that have come to infuse this term. The global repeats the claims of the universal male subject; unlocatable, a global outlook can seem to speak from everywhere, from a god's eye view. An implicit sexualization also runs through much of the literature of globalization. Gibson-Graham[3] for instance, argues that the metaphorics of rape and victimization permeate descriptions of globalization. Economics appears to trump all other factors and to define the global as little more than a network of markets and producers. Capitalism is figured as one cohesive, overpowering economic system. Women in particular are too often represented as passive objects of impersonal and unstoppable economic forces, either coerced to migrate or confined within the local scale, mired in their bodies and familial relations.

If discourses of globalization often rely on the same old stereotypes of masculinity and femininity, if the so-called global village, including global feminisms, often yields familiar categories of us and them, powerful and powerless, simply invoking the category of "the intimate" cannot unsettle these entrenched positions. Intimacy, after all, is equally caught up in relations of power, violence, and inequality and cannot stand as a fount of authenticity, caring, and egalitarianism. Svetlana Boym writes that intimacy, "may be protected, manipulated, or besieged by the state, framed by art, embellished by memory, or estranged by critique."[4] Intimacy does not reside solely in the private sphere; it is infused with worldliness. Nor is it purely personal: intimacy takes on specific political, social, and cultural meanings in different contexts. Feminists need to be skeptical of the ways in which the global and the intimate cross-pollinate in the new world order: from policy makers who deploy the language of idealized heterosexuality in the service of universalizing a single archetype of family; to the production of intimacy as a pageant for international media consumption; to the ways in which intimate tokens of gender identity, such as wearing "the veil," can rapidly metamorphose into potent symbols of cultural change and clash.

The pairing of the global and the intimate also asks us to consider how national visions of the global, particularly those located in the global north, tend to exclude anything that does not align with assumptions of northern hegemony and centrality. When you come down to the level of the intimate, the level of personal contact and exchange, individual differences and similarities become visible. This level of engagement forces the recognition of, for instance, the existence of cosmopolitan cities outside of the north; the presence of agency among so-called victims of globalization; the tendency to position middle-class Anglo women as modal subjects and to assume that all Anglo- and non-Anglo women share the same values and desires; and the uneven distribution of ethics and responsibilities worldwide.

In this introduction, we set forth, in a necessarily selective fashion, some of the major feminist approaches to both the intimate and the global and indicate more specific possibilities for how they might be productively read through and against each other. We live in an era in which the world continues to shrink, in which we are increasingly aware of how small gestures can generate wide ripples, and in which international forces explicitly determine our everyday lives. What Adrienne Rich wrote in 1984 seems newly urgent today: "Tribal loyalties aside and even if nation-states are now just pretexts used by multinational conglomerates to serve their interests, I need to understand how a place on the map is also a place in history within which . . . I am created and trying to create."[5] To inhabit a place on the map while simultaneously occupying what Rich called "the geography closest in—the body,"[6] to unsettle fixed perspectives, and to create practical and intelligent frameworks for a twenty-first-century feminist praxis: these are some of the agendas for this book. It is vital, at the outset, to acknowledge that both the global and the intimate have a complex theoretical history of feminist usage. We begin, then, by considering the ways in which women's and gender studies have approached both the global and the intimate, separately and together.

THE INTIMATE

"Intimate," from the Latin *intimus*, ("innermost"), is a word that invokes a cluster of related ideas: privacy, familiarity, love, sex, informality, and personal connection. Intimacy suggests something hidden away from the larger world, apparent only to the one or few on the inside. It refers to that which is walled off from the public sphere, from governance and regulation, from oversight. Intimacy has been traditionally associated with the feminine—and, not coincidentally, has sometimes been sidelined in scholarly inquiries. Feminist approaches to the intimate have sought to redress this exclusion and have distinguished within the sphere of intimacy a number of rubrics, prominently including feeling and affect; attachments to friends, families, and lovers; and the personal. Looking more closely at the development of this work can help us to see the diverse ways in which the discourse of intimacy can be connected to that of the global.

Feminist work on feeling has sought to place emotional life on the agenda of scholars, to understand feelings as important not just for psychologists but for fields of study as widely ranging as geography, political science, literature, anthropology, and economics. Often understood as the shifty and loose cousin of reason, emotion has been valued by both feminist and queer theorists not in spite of its variability and irrationality but because of it. This area of feminist inquiry carefully walks a line between investigating a central and vital aspect of human life and avoiding the reductive historical association of women with emotion. Attending to feeling can bring a valuable dimension to the subject of study, but it can also redound in interesting ways through a consideration of the feelings of the one doing the studying.

For many feminists emotion can be a potent analytic tool for discerning social injustices. Feminist scholars have been particularly attentive to what Sianne Ngai terms "ugly feelings,"[7] negative emotions that can be seen as expressions of thwarted agency, structural conflicts between women, or, in the most optimistic expressions, conduits to the raising of consciousness. In this context, anger has been of special importance. Audre Lorde writes, "The angers of women can transform difference through insight into power. For anger between peers births change, not destruction, and the discomfort and sense of loss it often causes is not fatal, but a sign of growth."[8] Anger becomes not just a subject for study by psychologists and a destructive or ego-driven force, but an agent of positive transformation. Implicit in Lorde's statement is the idea that the researcher/activist/writer does not stand outside the interplay of feelings in human exchange but is a part of the scene she is trying to understand and impact. Though researchers might prefer to see themselves as detached observers, everyone, to put it most plainly, has feelings.

Feminist scholars are hardly unified in their assessments of the politics of feeling. Wendy Brown, for instance, has argued that in its recourse to pain and reactive anger, feminism risks propounding a politics of *ressentiment*.[9] For Lauren Berlant, the widespread assumption that women share what she terms an "emotional generality" denies the fundamental diversity of women's histo-ries.[10] Others have seen more open-ended possibilities. For Ngai, even ugly feel-ings are more than reactionary on the one hand or libratory on the other; they are critically productive in a range of ways that stem from their overdetermi-nation and ambivalence.[11] In a global context, these insights make us both wary of and interested in the ways in which collective emotion can be mobilized (or manipulated) for a political agenda, whether it takes the form of liberal bene-volence, fear and disgust, or guarded optimism.

In recent studies across the academy, discussion of intimate feelings has ex-panded from emotion into the realm of affect—that is, into the body (including various nonorganic bodies and technologies that produce affective bodily capacities), as well as the mind. This shift is anchored in feminist thought. Michael Hardt, for instance, locates the precursors of the widespread interest in affect in a feminist focus on the body and explorations of emotion within queer theory.[12] His influential term *affective labor* describes embodied labor "that produces or manipulates affects such as a feeling of ease, well-being, satisfaction, excitement, or passion."[13] This is work that is most often done by women, or at least work that is most frequently associated with women's traditional capacities as caregivers, soothers, sex workers, or amanuenses. What Patricia Clough calls "the affective turn" in contemporary scholarship both grounds emotional life in the body and gestures toward something excessive to both emotions and the body: "affect constitutes a nonlinear complexity out of which the narration of conscious states such as emotion are subtracted, but always with a 'never-to-be-conscious autonomic remainder.'"[14] Affect theory challenges feminists to think beyond the body and emotions in a globalized frame; it speaks to the grounding and ungrounding of emotional life in relationships, both human and beyond, gesturing to and inculcating the experience of planetwide interconnection.

By enlarging on the terrain of emotion, affect slides into another convention-ally feminine province, that of relationality. Eve Sedgwick's work takes as its premise that affect does not reside within the discrete contours of the body but circulates among perceiving subjects (and objects). Although Sedgwick is pri-marily identified as a queer theorist, the continuity she describes between affect and attachment is also integral to feminist thought. Specifically, for Sedgwick the realm of feeling bleeds into that of touch, connected through the perceiving subject. "If texture and affect, touching and feeling seem to belong together, then, it is not because they share a particular delicacy of scale. . . . What they have in common is that *at whatever scale they are attended to,* both are irreduc-ibly phenomenological."[15] For Sedgwick, affect slides into physical connection just as the verb "to feel" shifts from intransitive to transitive. The connection between affect and interpersonal exchange forms an important nodal point in the geography of intimacy.

Sedgwick's work points up the continuity between two different aspects of feminist approaches to the intimate: feeling on the one hand and attachment on the other. Feeling is affect, but it is also feeling *something*; the perceiving subject is not an island but deeply and multiply connected to the world around her. Feminist scholarship on attachment tends to depict the subject as relational and critiques the masculinist bias in monadic models of subjectivity. As opposed to the isolated Cartesian *res cognitans*, for whom only his own interiority is real, the feminist subject is constituted through and by attachment. Models of this relational subject devised by psychologists offer a range of different ways to model the attached self and differentiate feminine developmental trajectories from masculine ones. The influential early work of Nancy Chodorow, for instance, asserts that girls "experience themselves as less differentiated than boys, as more continuous with and related to the external object-world."[16] Chodorow's image of a continuum suggests that we need not choose between a male isolate self and a female relational one but rather imagine differing degrees or kinds of attachments, determined not only by gender but by life experience, age, sexuality, and other elements. Carol Gilligan, building on Chodorow's work, argues that one extension of the relational self is an ethics motivated by concern for relationships and caring toward others.[17] More recently, Judith Butler[18] has drawn on Adriana Cavarero's[19] feminist grounding of selfhood in reciprocal exposure, dependence, and vulnerability to theorize a global relational ethics.

Attachment is a central rubric in feminist approaches to intimacy in part because it constellates a number of key ideas—relationships to others and to objects; sensory data and experiences of contact; bonds of family, desire, or affinity; and the commodification of all of these. Sara Ahmed writes, "The focus on attachments as crucial to queer and feminist politics is itself a sign that transformation is not about transcendence: emotions are 'sticky,' and even when we challenge our investments, we might get stuck. There is hope, of course, as things can get unstuck."[20] Intimate attachments, as Ahmed explains, can both inhibit and foster growth. This is the double bind of feminist theories of attachment: even as we claim and valorize women's relational identities, we critique the cultures of sentimentality when attachments become ties that bind and limit rather than strengthen. Feminist theorists often emphasize the ways in which attachments guide individual choices, values, and desires, but they have also been skeptical of the limitations attachments can confer. As early as the work of

Simone de Beauvoir, feminists have argued that emotional attachments can even cause us to feel loyal to conditions that enable our subordination and render women complicit in their own oppression. From de Beauvoir to Ahmed, attachment is a pivot point: a source of strength and self-definition, but also a boundary or sticking place.

Feminists have long recognized that intimate, sexual, familial, and other types of attachments are more than personal or private affairs; social, economic, and political worlds are built around personal attachments. Insofar as a capacity for empathy and compassion is one defining characteristic of the modern liberal bourgeois subject, it is our ability to form intimate relationships that places us in civil society. Elizabeth Povinelli traces a genealogy of critical thinking about the significance of the intimate couple in liberal thought and practice. "The intimate couple," she writes, "is a key transfer point between, on the one hand, liberal imaginaries of contractual economics, politics, and sociality and, on the other, liberal forms of power in the contemporary world."[21] Modern liberal forms of love and kinship, which emphasize individual freedom and choice, are anchored in a particular conception of the subject, the body and its extension in the world—in particular, a body and person that is bounded and self-sovereign. Attachment, popularly conceived as an individual commitment, is more usefully understood as a social practice with tremendous political freight. As Povinelli goes on to say, "Love is not merely an interpersonal event, nor is it merely the site at which politics has its effects. Love is a political event."[22] To understand love as a political event is to elaborate yet another dimension of the personal as political; it is to understand that the legal regulation of love through, for instance, marital and reproductive laws reflects the constitutive role of intimate attachments in the formation of nation-states—and vice versa.

The personal is not only an area of investigation but also a method or approach to scholarship, something like a style. Feminist theorists have made the personal an important basis for thinking about intimacy, from cleaving to what Nancy K. Miller identified as "feminist theory's original emphasis on the analysis of the personal."[23] For Miller, maintaining a personal voice or autobiographical presence in scholarly writing is a strategy that allows the author to reveal and dismantle the scrim of critical authority, to take apart the scholarly "we" that quietly creates the appearance of consensus and homogeneity by universalizing a necessarily personal perspective. It is a method, in other words, that can be used to disrupt power structures that construe themselves as impersonal and purely objective. In the context of globalization, the personal voice can function not only to reassert the primacy of identity but also as a discursive strategy for analyzing continuities and discontinuities among far-flung social worlds. For those of us who as teachers have pushed our students to adopt the first person "I" in scholarly writing, rather than the passive voice or the mysterious undefined "we," the personal voice is a way to ensure identification between the subject who writes and the subject being written about; it is a way to heighten the sense of responsibility for what is claimed on the page.

If the scholarly use of personal voice can transform modes of communication, it can also do important epistemological work. The same idea of locating the speaker in a particular nexus of beliefs and attributes can be recast as a means of demonstrating the inevitable partiality of the making of knowledge. Feminist standpoint theory, elaborated by Sandra Harding, Patricia Hill Collins, Nancy Hartsock, and others, asserts that all subjects have a standpoint, constructed in part by their social grouping, that influences how they understand the world.[24] In Donna Haraway's influential recasting of standpoint theory as "situated knowledges," she acknowledges that there can be no single feminist standpoint given the infinitely diverse and fractured perspectives women possess.[25] The idea of situated knowledges emphasizes locationality and promotes what Haraway calls "an epistemology and politics of engaged, accountable positioning"[26] that calls attention to the interplay between knowledge making and place. It recasts epistemological practices as intimate, subjective, and engaged, rather than impersonal and universal.

Though the identification of the person behind the politics in the way that situated knowledges mandate can be a feminist practice, this iterative style of describing identity by sorting into standard categories with standard choices (male/female, straight/gay, white/black, etc.) has been critiqued by feminists as static, reductive, and lacking a basis for solidarity or coalition. Cindi Katz has criticized a tendency for feminists to narrow the significance of situated knowledges to subjectivity and social location: "a space of zero dimensions . . . from which materiality is largely evacuated."[27] Katz expands the utility of the concept of situated knowledges in a globalizing world. She argues that knowledge should be expressed in terms that are material and site-specific, and that communicate the speaker's affiliations, and it is this highly specified epistemological approach, rather than a laundry list of attributes, that opens up more meaningful grounds for feminist solidarity. In thinking about the feminist personal, in other words, we can draw on theories of globalization to define the personal as encompassing more than translocational and relatively portable identity categories; we seek to understand the personal as closely bound to the practices and politics of place.

The personal, in specifying the individual on a number of levels, has a way of returning us to our bodies, to difference, to identity, to strength and fallibility, to ability and disability, to striving and limitation. The body can sometimes appear as mere housing for the brain, but it has enormous value and centrality for gender theorists. It may seem ironic that feminist theories of even the socially constructed gendered body can and frequently do reside at a high level of abstraction. When we invoke the idea of the intimate body, we set aside philosophical generalizations in favor of lived materiality: the body's history, recorded in freckles, scars, and creases; its preferences and pleasures; its surface appearance, influenced by fashion, fitness, function, or fantasy. The intimate body is not only differentiated, it is particularized.

Some French feminists writing in the 1970 and 1980s—including Luce Irigaray, Helene Cixous, and Monique Wittig—argued that feminist theory must both expose the ways in which women's bodies have been understood as lacking, castrated, or other and simultaneously craft an alternative tradition that takes gender difference (and other differences) as a grounding condition. More recently, feminists have endeavored to plot a course between the discursively constituted body described by Judith Butler on the one hand, and the material body that becomes visible through the social burdens that can attach to femaleness on the other (as seen, for instance, in Susan Bordo's work[28] on women's eating disorders). The body, as Gayatri Gopinath argues, can be and has been injured by racism and colonialism, but the body can also be a site of transformation and self-determination: "Queer diasporic cultural forms and practices point to submerged histories of racist and colonialist violence that continue to resonate in the present and that make themselves felt through bodily desire. It is through the queer diasporic body that these histories are brought into the present; it is also through the queer diasporic body that their legacies are imaginatively contested and transformed."[29] In Gopinath's understanding, the body is simultaneously global and intimate, interpreted through its relation to the history of global migrations and conflicts. As Art Spiegelman expresses it, individuals "bleed history"; they manifest through their bodies the processes of history.[30] The body, injured imaginatively and actually, testifies to historical violence, but it also is a site for resistance to coercive and deforming forces and a place for self-actualization or, at the least, a place from which to negotiate with social norms.

However, as Povinelli and Chauncey have argued, the body and subjectivity are absent from much of the literature on globalization: "A troubling aspect of the literature on globalization is its tendency to read social life off external social forms—flows, circuits, circulations of people, capital, and culture—without any model of subjective mediation . . . as if . . . an accurate map of the space and time of post-Fordist accumulation could provide an accurate map of the subject and her embodiment and desires."[31] The intimate forces our attention on a materialized understanding of the body when we theorize on a global scale. Is it possible to theorize global processes while remaining attentive to the pleasures and travails of individual embodiment? How might we find ways to hold on to emotion, attachment, the personal, and the body when we move into a more expansive engagement with the world? How, in other words, can we find the intimate in the global?

THE GLOBAL

"The global" typically evokes a historical conjuncture of economic, regulatory, and cultural forces that are called globalization or global restructuring. Some identifying features are these: deregulation of markets, privatization of services, flexible production, structural adjustment, a proliferation of governmental bodies alongside and beyond the nation state, networked connectivity, increased spatial mobility, transnational organizing, complex and creative intermingling of cultural forms, and the deep burrowing of capital accumulation into the body and circuits of affect. Women and girls have been at the centre of these many transformations and dislocations. As C. George Caffentzis wrote in 1993 of World Bank strategies to manage the debt crisis: "The African body, especially the female body, has been attacked."[32] Similarly, Chandra Mohanty writes, "It is especially on the bodies and lives of women and girls from the Third World/South—the Two-Thirds World—that global capitalism writes its script."[33]

Feminist engagements with these processes are vast and diverse; in fact, a hallmark of feminist scholarship is a certain skepticism about large, simplifying narratives of globalization. As Kathleen Stewart explains, the terms globalization, neoliberalism, and advanced capitalism cannot begin to describe the current situation and individuals' life circumstances.[34] Feminist engagement with the global has been marked by persistent critique, not only of mainstream global processes and discourses of globalization but also critique of Western feminists' fascination with and representations of women in the global south. We will review some of these criticisms and then consider the centrality of intimacy for the way that feminists have sought to think about global processes and their place within them, including the potential for feminist solidarity around the globe.

Feminist theorists have long noted the absence of gender as an explicitly theorized category and object of empirical investigation in both mainstream and more critical approaches to globalization. As Mary Hawkesworth notes, "While . . . authoritative analyses of globalization disagree about many things . . . they converge on one point: the near total absence of any reference to women or to feminism."[35] Gendered and sexualized metaphors of intimacy nonetheless undergird globalization discourses, both popular and scholarly, in unacknowledged and consequential ways. Charlotte Hooper has traced the implicit masculinities celebrated in popular discourse on globalization.[36] Examining *The Economist* through the 1990s, she notes that, even though the "new man" of global restructuring may be less formal and more technocratic than his forebears, as an aggressive risk taker pursuing capitalist adventures across the globe, he remains a resolutely masculine figure. The economy appears to trump all other (feminized) factors, and capitalism is figured as one cohesive, globally extensive system that can "have its way" with any local place it may seek to enter. By the same token, metaphors of the mismanaged body run through discourses of economic structural adjustment; Patricia Price describes how failing third-world economies are typically subjected to regimes of "belt tightening" and "fat cutting."[37] Unexamined, these metaphors of the slovenly or sexualized body nonetheless do considerable work, that of naturalizing and legitimating the seeming inevitability of global restructuring: the image of the fat, poorly managed body legitimates the massive privatization of state services; a victimized local economy appears to have no agency to stop her aggressor.

Feminists also express concern about the ways in which the intimate nuclear family, the protection of women and children, and some of the triumphs of second-wave feminism have been redeployed to both restrict the reach of ethical obligation and lend moral authority to global economic restructuring, increased securitization of national borders, policing within and beyond the nation, and military aggression worldwide. On the one hand, the family model, in particular the relation between parent and child, dominates and restricts our thinking about care; we tend to associate care with proximity, and we have few conceptual tools for imagining geographies of care beyond the national community. On the other, the model of the family is all too easily redeployed internationally; women and children are particularly effective vectors for sentimental politics, which can fuel both ethnic nationalisms and an abstracted, liberal human rights regime. Feminists have worried about the ways that their scholarship can be taken to support one of the most dangerous geographical imaginings of our time: the construction of a security state that justifies itself through its benevolent protection of women and children. Nancy Fraser is troubled as well by the "elective affinities" between some of the powerful critiques made by second-wave feminists, in particular of state bureaucracies and the family wage, and of neoliberalism and emergent forms of capitalism.[38] The clawback of the welfare state, the attendant devolution of service provision to nongovernmental organizations, and the expansion of women's waged employment in casualized, low-paying jobs since the 1970s may bear only a perverse resemblance to the emancipatory ideals of second-wave feminism, but the latter nonetheless endow these trends with a higher moral—and legitimating—significance.

So too, queer activists and theorists never imagined that their successes would be redeployed to harden the borders of the global north against Muslim migrants. And yet tolerance of sexual diversity in now taken in many counties in the global north as a measure of progressive Western liberalism (and the implicit backwardness of Islamic societies where such tolerance is not exhibited), a development that Jasbir Puar identifies as homonationalism.[39] In the Netherlands, for instance, the suitability of a subset of migrants (largely from non-Euro-American countries) for citizenship has been tested by responses to a photograph of two men kissing; in South Africa, tolerance of sexual diversity has been used to rebrand as progressive the postapartheid state. Reflecting on "the cunning of history" and the capacity for capitalism and hegemonic power blocs to remake themselves through a creative reworking of critique, including feminist critique, these and many other theorists insist on our responsibility to deliver nuanced, complex portrayals of worlds beyond the global north, worlds full of contradictions, agency, and struggle, and to recuperate realms of intimate life that are beyond but inextricably bound up with processes of capital accumulation.

To some extent, feminists have been no happier with many of their own efforts to come to grips with global processes. A number of queer theorists, for instance, have criticized the internationalization of Euro-American gay sexual identity as yet another repetition of colonial narratives of development and progress. As early as 1991, Rey Chow[40] criticized the tendency within Western feminism for white women to claim for themselves the position of investigator while casting non-Western women as alternatively victims of patriarchy or encased within "the culture garden" of another time. She noted a tendency to employ different methodological and theoretical approaches to the lives of Western and non-Western women, with Western women interpreted through sophisticated theories of intimate subjectivity (for example, psychoanalysis) and depth models of interiority, and non-Western women viewed with a less complicated subjectivity. In the case of the latter, the focus more often has been on the brute facts of their existence and the materiality of their lives.

Western women are typically seen as having complicated and contradictory subjectivities and desires, non-Western women only bodies and needs. Western feminists often unwittingly work within narratives of progress (with non-Western women presumed to experience more persistent and violent forms of patriarchy), and they essentialize and universalize particular conceptions of care, human rights, and models of (typically secular) gender equality. As one example of this, in an attempt to theorize the maternal beliefs and practices of mothers living in the shanty town of Alto do Cruzeiro, Brazil, where mothers' detachment from infants they consider to be doomed sometimes leads to a fatal neglect of these infants, Nancy Scheper-Hughes criticized the extent to which influential feminist theories on "maternal thinking," "feminine personality," and "womanly ethos" are rooted in the assumptions and values of the modern Western bourgeois family and "do violence" to alternative moral frameworks.[41] She reports that when she first wrote in 1984 about this "pragmatics of motherhood," attentive to the ways in which political and economic realities shape maternal thought and practice in particular places, she was criticized by other feminists for her lapse in sisterly solidarity. Some twenty-five years later, we might hope that universalizing a single maternal script across varying societies in different places is the more likely target of criticism.

And yet there is reason to be cautious about this expectation. In 1988 Chandra Mohanty, examining the entire list of books in the Zed Press Women in the Third World series, argued that Western feminists constructed third world women as homogeneous, powerless victims, wholly determined by their material circumstances.[42] If Western feminism was not a coherent body of theory, it was, she argued, coherent in its effects, in large part because "the West" was the inevitable norm against which all other peoples and places were judged (and found wanting). Writing some twenty years later, she remains concerned that feminist theorists continue to construct a cast of overly simplified feminine characters who are easily scripted into a gendered (and deeply masculinist) reading of globalization. "Clearly," Mohanty writes, "there is the ubiquitous global teenage girl factory worker, the domestic worker, and the sex worker. There is also the migrant/immigrant service worker, the refugee, the victim of war crimes, the woman-of-color prisoner who happens to be a mother and drug user, the consumer-housewife, and so on."[43]

In Mohanty's view, feminist scholars still tend to map the victims of globalization onto stereotyped locations in the global south, while making images of empowered women most visible in more privileged geographical locations. "The concern here," she writes, "is with whose agency is being colonized and who is privileged in these pedagogies and scholarship."[44] Spivak, too, notes a tendency for feminists in the global north to absorb processes of globalization into their own experiences. She has criticized the tendency to equate globalization with urbanization, migrancy, and diaspora, and to ignore rural populations, especially those in the global south.[45] Moreover, she argues that scholars often cast urban diasporic communities within familiar frameworks of subjectivity and intimacy—namely, the gender-race-class relations of the "receiving" country. Such analyses "remain narcissistic, question-begging" because they return readers to themselves and their own "predicament" of a multicultural society.[46] They treat the migrant as "an effectively historyless object of intellectual and political activism," thereby simplifying migrant subjectivities,[47] and reasserting the centrality of the metropolis in the global north and the irrelevance of all places and social relations of intimacy and attachment that lie outside it.

Intimacy and attachment are nonetheless being rethought in creative ways, and new globalized technologies are likely at the heart of some of this, especially the turn to affect and the fascination with the incoherence of bodies and identities, de- and reterritorialized as particles, intensities, and forces in body-technology assemblages of sensation that flow between and across human bodies, and inanimate and animate objects. Amit Rai writes of "qualitatively new (non) human bodies in the context of changing media assemblages";[48] Clough of "new configurations of bodies, technologies, matter."[49] This may take the form of cyborg assemblages of humans and machines, or intimate interminglings of particles of matter that destabilize assumed boundaries between humans, animals, and inanimate objects, along with attendant geographies of containment. Povinelli and Chauncey argue that globalization (and its attendant technologies) scrambles "the commonsense referents of the proximate and the intimate, the subject and her space and time and being, and thus her forms and practices of desire."[50] Information technologies, ubiquitous air travel (for some), inexpensive phone rates: these all throw into question the spatiality of intimacy and facilitate the proliferation of styles and forms of intimacy—for example, transnational families, international adoptions, sex tourism, and international migrant domestic labor. That so many (although certainly not all) of these new forms of intimacy emerge within patterns of global uneven development and involve the deep exploitation of women, often women from the global south, does, however, prompt us to ask whether feminism has a critical role to play in qualifying what can seem like a breathless celebration of body-technology assemblages by those who write within the affective turn.

Despite and possibly because of the persistent critique within feminism of a tendency to ignore, simplify, and flatten the lives of women in the global south, there is now a large body of feminist scholarship that teases out the nuances of global restructuring: intimacy, emotions, affect, and attachment are central to these narratives of constrained agency, complexity, and contradiction. If the teenage girl factory worker, the domestic worker, and the sex worker are ubiquitous, as Mohanty notes, it would be wrong to suggest that these figures are inevitably or even dominantly treated in hackneyed ways. Within the vast literature on young women's employment in global factories, low-end manufacturing jobs, and export or industrial-processing zones, for instance, there is, to be sure, a persistent emphasis on the ways that young women are easily scripted into this work through traditionally feminine characteristics of dexterity, patience, and conscientious attentiveness to detail. These factories typically depend on gendered and intimate forms of labor control that center on women's bodies and sexual harassment—for instance, physical confinement during working hours, strict regulation of trips to the toilet, and body searches to regulate pregnancy and verify requests for menstrual leave.

But within such contexts there is room for agency. It might take the form of what Pun Ngai has termed "a minor genre of resistance,"[51] literally on and through the intimate spaces of the body. In her 1987 ethnography of young women factory workers in Malaysia, for instance, Aihwa Ong documented the widespread incidence of spirit possession among those working in multinational factories, so extreme that it caused several factories to close for days at a time.[52] In the global factories that Pun studied in the special economic zone of Shenzhen China, women also live their discontent within the intimate spaces of their bodies, but in the more prosaic forms of menstrual and back pain, and fainting.

This raises pressing questions about the circumstances under which women's resistance moves from the minor genre of bodily complaint to more collective and overtly politicized forms of organizing. Highlighting the significance of the proximate and intimate, in a review of recent monographs on women workers in global factories Patricia Fernández-Kelly concludes that collective organizing is possible when women have networks of local attachments and intimate relations of support.[53] If this is so, the explosion of temporary migrant work worldwide is a particularly worrying trend.

We might, however, think harder about the intimate to stretch our imaginations about political possibility within what appears to be a forbidding global economy. Feminists have long argued that social cooperation, social reproduction, attachment, feeling, the personal, and the private are inseparable from the economy, and that the feminization of these increasingly commodified areas of life in capitalist societies devalues them in ways that are critical for the production of economic value. But the personal, the intimate, and social cooperation and attachment may also provide the ground for new or different political opportunities. Aihwa Ong, for instance, argues that NGOs in Asia are finding ways other than citizenship and the language of rights to revalue the politically excluded.[54] In the case of migrant domestic workers, the at-risk body of the female life-giver and life-nurturer can be used, Ong argues, more effectively than a language of abstract legal rights to appeal to employers' Asian family values, and to elicit their sympathy for domestic workers' bodily needs: their need for a day of rest, reduced overtime hours, and adequate health care. Neferti Tadiar also searches for possibilities beyond citizenship, rights, and resistance predicated on national or territorial identity or class struggle.[55] We miss other social logics and cultural capacities, she argues, if we reduce social and political experience to these categories.[56] And although she recognizes the limitations of a discourse of suffering and familial sacrifice, Tadiar posits (divine) sorrow as "a placeholder—a prayer, perhaps—for forms of political agency and notions of community beyond our prevailing notions of politics"[57] and social identity. Although we might hesitate about redeploying a moral vernacular of familial paternalism, as Ong does, or following Tadiar in finding emancipatory potential in collective practices of suffering, both suggestions are symptomatic of the search for political possibilities within the language and experiences of intimacy.

Feminists are also searching for new ways of thinking of global collective solidarity that both acknowledge and counter the fragmentation of feminism. Critiques of the Eurocentrism of feminist theory and practice, and appreciation of the particularity of women's experiences depending on race, class, and historical and geographical context (among other things) have thoroughly dispelled any assumptions about an inherent basis for identification among women: the fantasy of an unearned global sisterhood is well and truly dead. But aspirations to global solidarity and universal norms are not, and they involve new ways of thinking of both the global and the intimate. Locating this solidarity in the suffering or vulnerable body is one surprisingly popular avenue. Judith Butler pursues this idea (if more abstractly than Ong and Tadiar) as she searches for new grounds for global ethics and political obligation beyond the nation in the precariousness of life and literal, bodily vulnerability.[58] The recognition of precarity, she hopes, has the potential to establish affective circuits of global interdependency (which bear no relation to traditional notions of love and care

dependent on proximity and identification) that will introduce stronger normative commitments to equality and universal rights.

Another approach is to trace the ways that processes of globalization affect
women and girls, boys and men around the world in different and particular
ways. Cindi Katz develops the metaphor of counter-topography as a way of
imagining this.[59] Topography, she writes, is "the accurate and detailed description of any locality,"[60] the three-dimensionality of which is produced by a patient layering of contour lines, one on top of the other. She imagines particular
processes—say the expansion of feminized service employment—as contour
lines of constant elevation, connecting places. But such processes are always situated within and altered by the specifics of their fully three-dimensional place,
in relation to the other "contour lines" that define the topography of that place.
Tracing counter-topographies involves the simultaneous labor of following contour lines across places and understanding how global processes are embedded
in particular places.

In an effort to undercut humanitarian and paternalistic perspectives emanating from the global north, a number of feminists have insisted on the need to
understand the punishing effects of global economic processes close at hand as
well as far away. That is, benevolent subjects of the global north need equally to
turn their attention to conditions close to home—for instance, by balancing concerns about women's exploitation in sweatshops in Indonesia with anti-racism and
living-wage campaigns in solidarity with local (often migrant) workers. Feminist solidarity emerges not through some abstract identification as a woman or
a feminist but through actual, unpredictable, sometimes angry or distrustful
intimate relations, through the hard work of communication and collaboration.
The current enthusiasm for sustained participatory and transnational feminist
collaborations, of which the Sangtins (see chapter 14 in this volume) are exemplary, reflects this commitment to building rather than assuming solidarity.

THE GLOBAL AND THE INTIMATE:
A GUIDE TO THIS BOOK

As we have set out to show, feminists implicitly have been thinking about the
global in the intimate and vice versa for some time. All of the authors in this
book were asked to write explicitly within this pairing: three (Fidecaro, Grewal,
and Smith) for a previous journal issue on this theme,[61] the rest for this collection. By placing the global and the intimate into near relation, we hope to forge

a distinctively feminist approach to pressing questions of transnational relations, economic development, global feminist mobilization, and intercultural exchange. Although we have organized the essays within the familiar convention of scale, moving from closest in—the body—to the global, all of the essays in this collection explicitly adopt a method that involves disrupting the very idea of scale, by sliding between global and intimate, weaving together these two different modes of feminist thought. The essays are written from many different geographical locations: Australia, Amsterdam, Canada, India, Japan, Mexico, Switzerland, Taiwan, and the United States; and from varying disciplinary locations (indeed, most are unclassifiable in the conventions of these terms). They cover a wide range of topics.

Despite this diversity, there are, of course, limitations in this volume, felt most acutely by the editors in its geographical reach. It bears the traces of our own situations in Canada and the United States. Our ambition for this volume was to struggle against our locations to assemble the greatest possible range in the backgrounds and locations among our contributors. In falling—perhaps inevitably—short of this goal, we nevertheless learned something about the insularity (a form of intimacy?) of our worlds. Our call for papers for the journal issue of *WSQ* that we edited under this theme was widely circulated, but the majority of responses came from familiar places and institutions. We were limited from the start by our requirement that all articles be written in English. And our own feminist networks asserted themselves, as we turned to friends and colleagues to contribute.

We did, however, find some ways of resisting the centripetal force of insularity. Although we required that all articles be submitted in English, we solicited and supported work from nonnative speakers. We encouraged our contributors to write in different modes—some theoretical, some critical, some personal, some travel narrative. We tried to provide geographical diversity, including not just writers from different birthplaces, but writers working in and out of different locations.

Through all of this diversity, the collection demonstrates the creative potential that comes from pairing the global and the intimate. The intimate brings us in close, to the tastes, smells, and touch of everyday life and the materiality of social existence. It is perhaps not surprising that this emphasis on materiality is clearest in the earlier papers more focused on the geographies closest in: for example, Elspeth Probyn writing about the intimacy of consuming oysters, taste attachments, and place; Agnese Fidecaro on the clutter of seed catalogues and lists of garden nurseries in Jamaica Kincaid's garden writing; Marisa Belausteguigoitia

Rius on the collective storytelling made possible through the concrete materiality of the spiral and mural painting in a women's prison in Mexico. But an insistence on materiality runs through much of the book—for instance, Marianne Hirsch's emphasis on the significance of objects for narrating memories of traumatic dispossession and return, gesturing toward a broader commitment to rematerializing the social.

The intimate directs us to an ethical stance toward the world—namely, an approach that neither simplifies nor stereotypes but is attentive to specific histories and geographies. If we can guard against universalizing particular structures of intimacy and resist any facile connection between women and the intimate, we can find in the idea of the intimate a language and an approach that can disrupt tendencies toward totalization that inevitably arise when we try to theorize on a planet-wide basis. Several of the papers elaborate critiques of universalizing particular structures of intimacy and the risks that this poses for feminist organizing. Examining the commodification and reception of a young Bosnian-Croat girl's diary in Europe and the United States, Sidonie Smith reviews the global workings of a sentimental politics that beckons the reader through assumptions of universal parenthood. Melissa Wright is concerned about the production of the mother-activist in northern Mexico, honed in relation to international human rights testimonial regimes, and the negative implications that this has had for local feminist alliances in northern Mexico and broader campaigns against violence against women. Likewise, Inderpal Grewal examines how the discourse of security moms in the United States works to legitimate a hardening of national borders that deflects attention from domestic violence against women.

By the same token, "thinking globally" can provide a necessary counterweight to traditional analyses of intimacy by calling attention to the unintended consequences of failing to place the local setting in a broader context or refusing to look at the threads that connect intimate practices to a larger world. For Huang and Li, this means paying attention to the fact that a beloved caregiver is only assimilable to her employer's family because economic realities have forced her to leave her own kinship circle. For Fidecaro, it means upending simple associations between gardening, domesticity, self-containment, and stability. For Grewal, it means scrutinizing carefully the ways in which familial values and ideals shape not only the home front but also international relations. The intimate can be nearsighted, self-important, and limiting; toggling the intimate with the global is, among other things, an ethical turn designed to unsettle certainty and create accountability, to break out of the intense self-directed or dyadic gaze to acknowledge the world beyond.

A recurring theme through the collection is the need to attend to the particular; indeed, Hirsch and Probyn, following in the tradition of scholars like Naomi Schor, identify attunement to detail as a defining characteristic of feminist inquiry.[62] We believe that an explicit pairing of the global and the intimate can create openings for promising geographical imaginations that build connections while insisting on the specificity of processes in distinctive places. The papers explore how to build connections across this particularity: Mieke Bal through a "performative universalism" that is constantly challenged by singularity; Wright through a politics that does not rest on familiarity; Hirsch through a methodology of connection rather than comparison. Building these connections across difference is no easy matter, and a willingness to embrace the intimate challenges of confrontation and collaboration is another defining characteristic of several of the contributions. In detailing their journey from a small writing and story-sharing collective of seven women to a peasant mass movement in a rural district of Uttar Pradesh, India, the Sangtins write of the challenges of tirelessly confronting the contradictions that persist, inevitably, in the complex politics of alliance work at transnational, national, and even local scales, and between academics and activists. Writing about her experiences of exposure to white women's stereotypes about Asian women, Min Jin Lee imaginatively restages some of these encounters with her Asian-American friends in tow, "not to have a rumble, but to ask questions and to tell our stories." She imagines that "we would have had a good cry and more laughs." Lee finds it easy to make emotional connections with her women friends who share a racial identification, much harder to say how she feels and finds common ground with women who are different. Intimacy doesn't come easy, yet it remains Lee's goal to be able to talk about "love with love" with "girls everywhere." Can an aspiration to global sisterhood bring emotional intimacy within reach?

Crying and laughing, it turns out, are part and parcel of feminist knowledge production and pedagogy. The Sangtins ask, "Which forms of knowledge and expertise remain unrepresented when tears do not find a space in intellectual conversations?" The logic and structure of the intimate creates—in fact, demands—such a space and suggests different modes of writing, different forms of address, and a rethinking of what our theories can act upon and what they are meant to do. As Fidecaro reminds us, the intimate is less easily objectifiable, possibly incommunicable—so as to provoke, even demand, experimentation. We solicited personal writing explicitly from Rachel Adams, Mikhal Dekel, Nancy K. Miller, and Min Jin Lee, but the personal creeps into many of the chapters, as authors fuse personal and academic writing, or tell of their experiences as teachers, or artists, or activists, or friends, or as lovers of oysters. As Mieke Bal puts it, the authors are searching for a methodology suitable to their object of study, and this takes different, often hybrid forms. To find emotional connection without insisting on familiarity, as Wright argues, is not easy; building coalitions remains hard work.

The intimate allows us to break out of established categories; it creates an opening to think something new. In her chapter in this volume, Ara Wilson argues that the turn to intimacy is bound up with a desire to break out of restrictive categories operating at a variety of scales: identity, kinship, the nation. It implies relationality but does not specify the form that this will take. For Probyn, it expands to include the human and nonhuman; for Adams, it takes shape through the Internet; for Bal, it emerges through the face and speaking across unbridgeable gaps. Fidecaro likewise argues that intimacy, with all of its associations with the unruly body, has a "reopening power that interrogates from inside all closed and enclosing structures that have to do with (self-) possession and property." Intimacy does not work within the same territorial or juridical logics as privacy, the local, or even the global. At the same time, the stain of what we call the global complicates and compromises intimacy in productive ways by opening it to histories of imperialism, national formation, global economic development, systematic humiliation and deprivation, and gender and sexual inequality. Joining the global and the intimate requires a constant attention to proportion, as well as to scale. It forces us to question what is big and what is small, what is important and what is inconsequential.

The work of our contributors demonstrates that the joint critical analysis of the global and the intimate is feminist work, work that can strengthen and extend feminism in our time. One of our aims in this volume is to show what a joint analysis of the global and the intimate can offer in diverse disciplines, including not only women's studies but geography, literature, anthropology, sociology, history, art history, and more. What we find most promising are the ways that some of our contributors have begun to build connections in the pages of this book. The Sangtins write of a trip to the United States and the meaningful conversations they began to have with migrant rights groups there; Belausteguigoitia writes of the conceptual connections that women prisoners in a Mexican prison began to make with prisoners in other parts of the world. Our hope is that the various chapters in this book can be put into conversation with each other, with some aspects of your world, and with all of our futures.

NOTES

1. Doreen B. Massey, *For Space*; Inderpal Grewal and Caren Kaplan, *Scattered Hegemonies*; Sallie A. Marston, John Paul Jones III, and Keith Woodward, "Human Geography without Scale"; J. K. Gibson-Graham, *The End of Capitalism (As We Knew It)*.
2. Marston et al., "Human Geography without Scale."
3. Gibson-Graham, *The End of Capitalism*.
4. Svetlana Boym, "On Diasporic Intimacy," 500.
5. Adrienne Rich, "Notes Toward a Politics of Location," 212.
6. Ibid.
7. Sianne Ngai, *Ugly Feelings*.
8. Audre Lorde, "The Uses of Anger," 131.
9. Wendy Brown, "Wounded Attachments," 394.
10. Lauren Berlant, *The Female Complaint*, 5.
11. Ngai, *Ugly Feelings*.
12. Michael Hardt, "Foreword: What Affects Are Good For," ix.
13. Michael Hardt, "Affective Labor," 96.
14. Patricia Clough, "Introduction," 2.
15. Eve Kosofsky Sedgwick, *Touching Feeling*, 21.
16. Nancy Chodorow, *The Reproduction of Mothering*, 167.
17. Carol Gilligan, *In a Different Voice*.
18. Judith Butler, *Giving an Account of Oneself*.
19. Adriana Cavarero, *Relating Narratives*.
20. Sara Ahmed, *The Cultural Politics of Emotion*, 16.
21. Elizabeth A. Povinelli, *The Empire of Love*, 17.
22. Ibid., 175–76.
23. Nancy K. Miller, *Getting Personal*, ix.
24. See, e.g., Sandra Harding, *The Science Question in Feminism*; Patricia Hill Collins, "The Social Construction of Black Feminist Thought"; Nancy C. M. Hartsock, *The Feminist Standpoint Revisited and Other Essays*.
25. Donna Haraway, "Situated Knowledges."
26. Ibid., 590.
27. Cindi Katz, "On the Grounds of Globalization," 1230.
28. Susan Bordo, *Unbearable Weight*.
29. Gayatri Gopinath, *Impossible Desire*, 4.
30. Art Spiegelman, *Maus I*.
31. Elizabeth A. Povinelli and George Chauncey, "Thinking Sexuality Transnationally," 445.
32. C. George Caffentzis, "The Fundamental Implications of the Debt Crisis for Social Reproduction in Africa," 31.
33. Chandra Talpade Mohanty, "'Under Western Eyes' Revisited," 514.
34. Kathleen Stewart, *Ordinary Affects*, 1.

35. Mary E. Hawkesworth, *Globalization and Feminist Activism*, 173, n. 1.
36. Charlotte Hooper, *Manly States*.
37. Patricia Price, "No Pain, No Gain."
38. Nancy Fraser, "Feminism, Capitalism, and the Cunning of History."
39. Jasbir K. Puar, *Terrorist Assemblages*.
40. Rey Chow, "Violence in the Other Country."
41. Nancy Scheper-Hughes, *Death Without Weeping*.
42. Chandra Talpade Mohanty, "Under Western Eyes."
43. Mohanty, "'Under Western Eyes' Revisited," 527.
44. Ibid., 528.
45. Jenny Sharpe and Gayatri Spivak, "A Conversation with Gayatri Charkravorty Spivak."
46. Gayatri Charkravorty Spivak, "Thinking Cultural Questions in 'Pure' Literary Terms," 335.
47. Ibid.
48. Amit Rai, *Untimely Bollywood*.
49. Clough, "Introduction," 2.
50. Povinelli and Chauncey, "Thinking Sexuality Transnationally," 443.
51. Pun Ngai, *Made in China*.
52. Aihwa Ong, *Spirits of Resistance and Capitalist Discipline*.
53. Patricia Fernández-Kelly, "The Global Assembly Line in the New Millenium."
54. Aiwha Ong, *Neoliberalism as Exception*.
55. Neferti Tadiar, *Things Fall Away*.
56. Ibid.
57. Ibid., 378.
58. Judith Butler, *Precarious Life*.
59. Katz, "On the Grounds."
60. *Oxford English Dictionary*, s.v. "topography," in Katz, "On the Grounds," 1214.
61. Geraldine Pratt and Victoria Rosner, eds., *The Global and the Intimate*.
62. Naomi Schor, *Reading in Detail*.

BIBLIOGRAPHY

Ahmed, Sara. *The Cultural Politics of Emotion*. New York: Routledge, 2004.

Berlant, Lauren. *The Female Complaint: The Unfinished Business of Sentimentality in American Culture*. Durham: Duke University Press, 2008.

Bordo, Susan. *Unbearable Weight: Feminism, Western Culture, and the Body*. Berkeley: University of California Press, 1993.

Boym, Svetlana. "On Diasporic Intimacy: Ilya Kabakov's Installations and Immigrant Homes." *Critical Inquiry* 24, no. 2 (Winter 1998): 498–524.

Brown, Wendy. "Wounded Attachments." *Political Theory* 21, no. 3 (1993): 390–410.

Butler, Judith. *Giving an Account of Oneself.* New York: Fordham University Press, 2005.

——. *Precarious Life: The Powers of Mourning and Violence.* London: Verso, 2004.

Caffentzis, C. George. "The Fundamental Implications of the Debt Crisis for Social Reproduction in Africa." In Mariarosa Dalla Costa and Giovanna F. Dalla Costa, eds., *Paying the Price: Women and the Politics of International Economic Strategy*, 15–41. London: Zed Books, 1995.

Cavarero, Adriana. *Relating Narratives: Storytelling and Selfhood.* London and New York: Routledge, 2000.

Chodorow, Nancy. *The Reproduction of Mothering*, 2nd ed. Berkeley: University of California Press, 1999.

Chow, Rey. "Violence in the Other Country: China as Crisis, Spectacle, and Woman." In Chandra Talpade Mohanty, Ann Russo, and Lourdes Torres, eds., *Third World Women and the Politics of Feminism*, 81–100. Bloomington: Indiana University Press, 1991.

Clough, Patricia. "Introduction." In Patricia Ticineto Clough with Jean Halley, eds., *The Affective Turn: Theorizing the Social*, 1–33. Durham: Duke University Press, 2007.

Collins, Patricia Hill. "The Social Construction of Black Feminist Thought." *Signs* 14, no. 4 (1989): 745–73.

Fernández-Kelly, Patricia. "The Global Assembly Line in the New Millenium: A Review Essay." *Signs* 32, no. 2 (2007): 509–21.

Fraser, Nancy. "Feminism, Capitalism, and the Cunning of History." *New Left Review* 56 (March–April 2009): 97–117.

Gibson-Graham, J. K. *The End of Capitalism (As We Knew It): A Feminist Critique of Political Economy.* Malden, MA: Blackwell, 1996.

Gilligan, Carol. *In a Different Voice.* Cambridge: Harvard University Press, 1982.

Gopinath, Gayatri. *Impossible Desires: Queer Diasporas and South Asian Public Cultures.* Durham and London: Duke University Press, 2005.

Grewal, Inderpal, and Caren Kaplan. *Scattered Hegemonies: Postmodernity and Transnational Feminist Practices.* Minneapolis and London: University of Minnesota Press, 1994.

Haraway, Donna. "Situated Knowledges: The Science Question in Feminism and the Privilege of a Partial Perspective." *Feminist Studies* 14 (1988): 579–99.

Harding, Sandra. *The Science Question in Feminism.* Ithaca and London: Cornell University Press, 1986.

Hardt, Michael. "Affective Labor," *boundary 2* 26, no. 2 (1999): 89–100.

——. "Foreword: What Affects Are Good For." In Patricia Ticineto Clough with Jean Halley, eds., *The Affective Turn: Theorizing the Social*, ix–xiii. Durham: Duke University Press, 2007.

Hartsock, Nancy C. M. *The Feminist Standpoint Revisited and Other Essays.* Boulder: Westview Press, 1998.

Hawkesworth, Mary E. *Globalization and Feminist Activism.* Lanham: Rowman and Littlefield, 2006.

Hooper, Charlotte. *Manly States: Masculinities, International Relations, and Gender Politics.* New York: Columbia University Press, 2001.

Katz, Cindi. "On the Grounds of Globalization: A Topography for Feminist Engagement." *Signs* 24, no. 4 (Summer 2001): 1213–28.

Lorde, Audre. "The Uses of Anger: Women Responding to Racism." In *Sister Outsider*, 124–33. Berkeley: Crossing Press, 1984.

Marston, Sallie A., John Paul Jones III, and Keith Woodward. "Human Geography Without Scale," *Transactions of the Institute of British Geographers* 30 (2005): 416–32.

Massey, Doreen B. *For Space*. London: Sage, 2005.

Miller, Nancy K. *Getting Personal: Feminist Occasions and Other Autobiographical Acts*. New York and London: Routledge, 1991.

Mohanty, Chandra Talpade. "Under Western Eyes: Feminist Scholarship and Colonial Discourses." *Feminist Review* 30 (1988): 61–88.

——. "'Under Western Eyes' Revisited: Feminist Solidarity Through Anticapitalist Struggles." *Signs* 28, no. 2 (2002): 499–535.

Ngai, Pun. *Made in China: Women Factory Workers in a Global Workplace*. Durham: Duke University Press, 2005.

Ngai, Sianne. *Ugly Feelings*. Cambridge and London: Harvard University Press, 2005.

Ong, Aiwha. *Neoliberalism as Exception: Mutations in Citizenship and Sovereignty*. Durham: Duke University Press, 2006.

——. *Spirits of Resistance and Capitalist Discipline: Factory Women in Malaysia*. Albany: State University of New York Press, 1987.

Povinelli, Elizabeth A. *The Empire of Love: Toward a Theory of Intimacy, Genealogy, and Carnality*. Durham: Duke University Press, 2006.

Povinelli, Elizabeth A., and George Chauncey. "Thinking Sexuality Transnationally: An Introduction." *GLQ* 5, no. 4 (1999): 439–50.

Pratt, Geraldine, and Victoria Rosner, eds. *The Global and the Intimate*. *WSQ* 34, nos. 1 and 2 (Spring/Summer 2006).

Price, Patricia. "No Pain, No Gain: Bordering the Hungry New World Order." *Environment and Planning D: Society and Space* 18 (2000): 91–110.

Puar, Jasbir K. *Terrorist Assemblages: Homonationalism in Queer Times*. Durham: Duke University Press, 2007.

Rai, Amit. *Untimely Bollywood: Globalization and India's New Media Assemblages*. Durham: Duke University Press, 2009.

Rich, Adrienne. "Notes Toward a Politics of Location." In *Blood, Bread and Poetry: Selected Prose 1979–1985*, 210–31. New York: Norton, 1986.

Scheper-Hughes, Nancy. *Death Without Weeping: The Violence of Everyday Life in Brazil*. Berkeley: University of California Press, 1992.

Schor, Naomi. *Reading in Detail: Aesthetics and the Feminine*. New York and London: Routledge, 2006.

Sedgwick, Eve Kosofsky. *Touching Feeling: Affect, Pedagogy, Performativity*. Durham and London: Duke University Press, 2003.

Sharpe, Jenny, and Gayatri Spivak. "A Conversation with Gayatri Charkravorty Spivak: Politics and the Imagination." *Signs* 28, no. 2 (Winter 2003): 609–24.

Spiegelman, Art. *Maus I: A Survivor's Tale: My Father Bleeds History*. New York: Pantheon, 1986.

Spivak, Gayatri Charkravorty. "Thinking Cultural Questions in 'Pure' Literary Terms." In Paul Gilroy, Lawrence Grossberg, and Angela McRobbie, eds., *Without Guarantees: In Honour of Stuart Hall*, 335–57. London and New York: Verso, 2000.

Stewart, Kathleen. *Ordinary Affects*. Durham: Duke University Press, 2007.

Tadiar, Nerferti. *Things Fall Away: Philippine Historical Experience and the Making of Globalization*. Durham: Duke University Press, 2009.

I

THE
ANATOMY
OF
INTIMACY

Bodies,

Feelings,

and the

Everyday

1

INTIMACY
A USEFUL CATEGORY OF TRANSNATIONAL
ANALYSIS

Ara Wilson

THE WORK OF INTIMACY

Recently, feminist and queer studies of global power have turned to the concept of intimacy both as a subject and as an analytic rubric. They pair intimacy with globalization, or with its predecessors, colonialism and imperialism, or with the umbrella concepts, modernization and capitalist modernity.[1]

I see three reasons for this global-intimate pairing. One reason is that global political economic conditions have profound effects on human relationships, notably by introducing and then altering sweeping divisions between realms deemed public and private. Second, as feminist and queer works insist, intimate life is not confined to the private sphere but plays a role in the presumably impersonal spheres of government and economy, which in turn regulate the intimate domain. The third reason that scholars are examining intimacy alongside globalization stems from their dissatisfaction with the established terms used to understand the relationship between these domains, alternately macro and micro, global and local, public and private.

This essay offers an overview of interpretations of intimate life in global capitalist modernity. It centers on critical approaches to the political economy of intimacy—frameworks that are designed to understand how patterns of intimacy occur in relation to social power. Accordingly, the essay emphasizes concrete descriptions that rework received interpretations of globalization, government, capitalism, and intimacy. The cases discussed include particular countries (the United States, Thailand, Australia) and studies of transnational processes that cross countries' borders.

The concept of intimacy captures deeply felt orientations and entrenched practices that make up what people consider to be their "personal" or "private" lives and their interior selves, and includes positively valued feelings like affection but also problematic feelings like fear or disgust. The works I discuss here use intimacy to describe modes of relatedness associated loosely with personal feelings or identifications, in contrast, at least officially, to formal interactions within governments, markets, or modern institutions. As readers will see, if they do not already, the meanings of intimacy in these discussions vary quite a bit. The meanings are not fixed and can often seem vague: *intimacy* is not a term of art in any field. Why, then, does this loose term hold appeal for global analysis? I reached for the term *intimacy* in my own ethnographic work on Thailand.[2] My mixed results led me to become interested in understanding how other scholars are using the term and why they find the rubric of intimacy productive.

My conclusion is that the term's very lack of fixity is part of its appeal. It allows scholars to produce descriptions of the world order that do not re-create but rather scrutinize concepts that have often unwittingly perpetuated the inequality produced by governments and capital. By not building on the inherited associations of concepts associated with intimacy—concepts like family—the rubric facilitates a nondeterministic, nonreductive exploration of structures of feeling,[3] public feelings,[4] and biopolitics in relation to globalizing contexts. Used critically, the concept of intimacy facilitates the simultaneous recognition of social patterns in relationships and ideological norms about relationships. The term "intimacy" offers an appealing rubric for interpretations that undo familiar connotations about "private" life by emphasizing its historical and social situation—for example, in the everyday effects of global modernity or the inner operations of social hierarchies. The essay explores the promise of "intimacy" as an analytical, not merely descriptive, term for critical scholarship on globalization.

Collectively, much of the critical work on intimacy shows how patterns in intimate life have changed with realigned boundaries of public and private in civic life, governments, commerce, and nuclear families. As an illustration of these shifts, the essay offers the example of gated communities and shopping malls in the United States. Then I look at ways that norms about intimacy are bound up with hierarchies of race, nation, and sexuality. These examples focus on Europe's former settler colonies that are now multicultural liberal nation-states and a substantial portion of what is called the first world. I then turn what is known as the third world or global south, emphasizing a transnational orientation to global/local relations that recognizes linkages across richer and poorer countries (or global north and south). The rubric of "intimate economies" provides one model for thinking about global intimacy in ways that avoid a top-down image of impersonal forces "penetrating" intimate life, understood as local. The final section of the essay extrapolates from these cases the key themes informing the study of global intimacies in order to outline the emerging use of *intimacy* as a critical analytical term.

GATED INTIMACY

> *Wherever the title of streets and parks may rest, they have immemorially been held in trust for the use of the public and, time out of mind, have been used for purposes of assembly, communicating thoughts between citizens, and discussing public questions. Such use of the streets and public places has, from ancient times, been a part of the privileges, immunities, rights and liberties of citizens.*
>
> —1939 U.S. SUPREME COURT DECISION

The critical intimacy scholarship I discuss here places intimacy in relation to global modernity, in particular to late-twentieth-century social shifts associated with transnational capitalism. The kinds of intimacy these works consider include but go beyond conjugal couples and nuclear families. In fact, they want to recognize a form of public intimacy, the intimacy of public assembly, the kind of relations that involve "discussing public questions," which the 1939 U.S. Supreme Court decision cited above used to argue for the right of union advocates to discuss labor rights in public. Critical explorations of global intimacies explore ideals about relatedness (particularly the ways in which race is involved in norms for intimate life); the infrastructure for intimacy; and ways people live out everyday relations.

One of the main sites for investigating modern intimacy is the United States, which surely is due to the solipsism of U.S. researchers and to U.S. global power, but which is also an understandable focus, given that the United States represents a frontier of capitalist social experimentation. The portrait that has emerged in American Studies is of privatized public intimacy. Understanding that public civic life is formed through connections that involve forms of intimacy, this work shows how changes over the past few decades have supplanted public intimacies with privatized commercial or domestic forms. (Indeed, the U.S. Supreme Court's 2010 decision in Citizens United *v.* Federal Election Commission, which expanded the rights of corporations to speech designed to influence elections, offers an illuminating contrast to the 1939 Court's understanding of free speech as a public good.) At the same time, valorized forms of intimacy—particularly the conjugal couple, usually heterosexual but now also homosexual—have become national symbols in ways that exclude other relationships and that reinforce racial or national hierarchies.

One concrete example of the new geography of intimacy in the United States is the rise of the gated community. The names of many of these planned communities suggest roots in aristocratic estates, conveyed by Anglo-Saxon words for woods (oaks), water (creek, falls, lake), or large houses (manor)—"names that whisper exclusivity," as *USA Today* puts it.[6] This fantastical heritage aside, these residential arrangements represent a new mode of creating community, one that numerically represents a significant proportion of American residences.[7] As Setha Low explains in her study of such a community, gated communities are "a response to transformations in the political economy of late-twentieth-century urban America." She summarizes the broad trends: "Globalization and economic restructuring also weaken existing social relations. . . . Social control mechanisms, such as the police and schools, are no longer seen as effective. This breakdown in local control threatens some neighborhood residents, and the gated residential community becomes a viable and socially acceptable option."[8]

Commentators agree that people's attraction to planned communities reflects frustration with the government's provisioning of security or services: in short, they seek solutions through private property rather than through public government. Such fears of lessened control, Low notes, are imagined through racial and class terms. Gated communities allow residents to use capitalist markets and property rights to construct a controlled mode of intimacy. Gated communities strictly regulate their space, limiting house colors, street parking, number and kinds of pets, or numbers of visitors. One woman, described in the report as a grandmother, violated policies by kissing a friend goodnight outside her house.[9] Gated communities thus restrict the behaviors of their members in ways that bear upon their relations with other people (and animals). They also restrict the freedoms of nonmembers. These regulations put pressure on contradictions between private contract, property rights, and individual rights that have led to legal challenges in court.[10]

Residence associations selectively substitute private services for public services. While some of these developments build from the ground up, most convert preexisting public roads and infrastructure into private spaces, using law to authorize this transformation. Gated communities, analysts suggest, replace public space with privatized spaces, yet still they do not entirely fund themselves: rather they draw on public resources, including fire departments or special education for children, or rely on lessened obligations to the region or nation in the form of tax breaks.[11] Some legal reasoning proposes that, because the residential associations wield power associated with the state—they are "virtual governments"[12]—they should be treated as statelike bodies in the law. The point here is that the changes to public allocations of security, education, land, and so forth, and the increasingly private versions of neighborhood life, represent crucial conditions for intimate life, broadly understood. The withdrawal from surrounding publics creates a separate space for local relations that are predicated on exclusivity and defined by private property. The intimacies that are prioritized in these secure residential formations are nuclear families, intraclass, and planned rather than serendipitous: they are also often racially homogenous, and often populated with white people (although a relatively high proportion of Latinos reside in gated communities as well).[13] The conditions that enable such a withdrawal, and the desire for it, are indicative of trends in modern American intimacy.

Shopping malls also privatize the public, while being subsidized by public funds. The new trend in shopping malls, the "lifestyle mall," recreates old-time shopping districts, with "pedestrian friendly streetscapes"[14] and pseudo village greens, usually in upscale suburbs. In Columbus, Ohio, the Easton Town Center is a large mall that incorporates luxury rental apartments. It offers a fountain for toddlers' enjoyment. The planners of Easton emphasize their "philosophy of 'place making,' or positively impacting communities through the creation of dynamic mixed-use town centers."[15] In architecture and imagery, lifestyle shopping malls and gated communities invoke the symbolism of older forms of intimacy that were created in the public spheres of towns. Upscale stores are located in one-story "shops" lining private lanes given classic street names, like Main Street. The intended impression is that Banana Republic or Pottery Barn stores have moved into preexisting spaces downtown that formerly housed the drug store and its soda fountain, the haberdashery, or the barbershop.

Images of old-fashioned publics signal nostalgia for a world that never truly existed as imagined and are not re-created in a commercial simulacra. The Streets at Southpoint, a shopping center in Durham, North Carolina, includes bronze sculptures in the stylized realism of Norman Rockwell depicting children in various activities: playing in the decorative fountain, selling newspapers, walking a dog, or climbing a lamppost—notably, activities that would be prohibited in those very spaces of the mall today. Mall design sentimentalizes public life while the way of life the malls enact—for example, dependence on automobiles and privatization of public land—eviscerates actual public commons. Posted signs list prohibited activities, which include political activities like campaigning[16] and in particular restrict the actions of teenagers with curfews and codes for dress or conduct. These prohibitions reflect the fact that this is private space where many rights otherwise guaranteed in public spaces can be limited. Malls and gated communities exemplify commercially produced modes of selective, privatized public intimacy. Why has this privatized intimacy, defined by consumption, replaced activities oriented to neighborhoods, solidarity among workmates, ethnicity, or political parties?

The political scientist Robert D. Putnam captures changes to local relationships with the memorable image of Americans "bowling alone."[17] Bowling leagues, once prevalent, have diminished: more people bowl solo in the lanes. This quotidian example, for Putnam, is symptomatic of broader changes in people's sense of relatedness in the United States. It conveys a widespread decrease in civic participation, whether in community centers, religious networks, or union membership. Putnam attributes these transformations to such features of modern life as residential patterns (the rise of suburbs) and commercial media (television). The erosion in public life has political consequences. It erodes a sense of group interest and dialogue about public life, substituting the discourse of national television shows or talk radio for conversations in the community. It erodes what can be thought of as public intimacy.

SETTLER INTIMACY

Gated communities, and their nostalgic constructions of family and neighborhood, illustrate a point that Elizabeth Povinelli, Lauren Berlant, and others have explored in depth: that in late-twentieth-century Western societies—Europe and its settler societies (Australia, Canada, South Africa, the United States), the evaluation of forms of intimacy is tied up with social inequalities. A valorization of the conjugal couple, in particular, operates symbolically through notions of modernity and nationality and pragmatically through law, economics, and popular culture. Public evaluations of proper forms of intimacy tend to involve racial and national associations, denigrating modes of relations that characterize first-world communities of color, indigenous populations, or non-Western societies. In this way, investments in certain forms of intimacy as emblems of liberal modernity perpetuate de facto, if not de jure, social inequality.

Moreover, as with the erosion of bowling leagues and other forms of public participation, consequential judgments about the worth of different modes of intimacy also erode the intimacy of the public sphere.

A cardinal example of the interconnection of race, nation, and intimacy is miscegenation law, an interconnection that was reformulated, not ended, with its dismantling in the United States by the Supreme Court decision in Loving *v.* Virginia. With the end of most formal regulations of individual consensual intimacy, evaluations of relationships remain bound up with racial and national inequality. Robyn Wiegman's essay "Intimate Publics: Race, Property, and Personhood" considers the ways that kinship relations are adjudicated in a U.S. context where biotechnology, formal racial equality, and liberal economics present a new terrain of family formation. Contract relations, which mediate "between the seemingly private world of personal affect, intimacy, and reproduction and the public realm of social exchange," take on greater weight in establishing legitimate kinship relations.[18] Reproductive technologies like surrogacy or in vitro fertilization rely on contracts to navigate new modes of reproduction and kinship, but at times the outcomes of reproductive technology are legally contested, particularly when they involve an interracial set of participants. Many court decisions betray "class, race, and heterosexual assumptions about 'proper' maternity and 'good' family that coalesce in the naturalization of the white patriarchal nuclear family as the state's normative ideal."[19] These contestations over kinship cannot be reduced to one axis of race or gender, Wiegman argues; rather she emphasizes a plurality, noting that decisions about kinship in complicated reproductive scenarios are informed by multiple histories of racialization (including histories of immigration and whiteness), sex/gender regimes, and economic operations (class, property, and contract).[20]

Also exploring discourses of relationships in American neoliberal modernity, Lauren Berlant's work offers a highly influential framework for critical approaches to intimacy. Berlant excavates interpretations of modern intimacy in the United States, particularly the privileged place of married heterosexuality in conceptions of America. She suggests that "a virulent form of revitalized national heterosexuality," one "that is complexly white and middle class,"[21] has become the anointed emblem of U.S. citizenship: "A nationalist ideology of marriage and the couple is now a central vehicle for the privatization of citizenship: first, via moralized issues around privacy, sex, and reproduction that serve as alibis for white racism and patriarchal power; but also in the discourse of a United States that is . . . an effect of the private citizen's acts."[22]

Berlant argues that such privileging privatizes civic life and erodes public intimacy. Narrating civic life through celebrations of married, reproductive, heterosexual intimate lives perpetuates "a fantasy that private life is the real in contrast to collective life."[23] Prioritizing family in this way therefore "supports disinvesting in cultural, collective forms of personhood while promoting an image of the legitimate, authentic individual situated in the spaces of intimate privacy."[24] The legitimate forms of intimacy are normative domestic life—mainly heterosexual, but in some areas homosexual as well—and/or bonding through shared references to Hollywood stars, television shows, or top-40 music. Civic life becomes reduced to the couple or the television viewer.

Elizabeth Povinelli's iconoclastic project *The Empire of Love* explores contrasting evaluations of intimacy in the United States and Australia. In these settler societies, she says, intimate relations are categorized according to their creation through social inheritance or individual choice, a criteria that also maps onto racial identities. Studying an intentional group known as radical faeries, who are associated with modern life, and Australian aborigines, who are associated with tradition, Povinelli shows how the assumptions about intimacy in settler states has material consequences.[25] Like Berlant, she sees whiteness and racialization as central to official differentiations of intimacy in these liberal multicultural states.

Echoing Michel Foucault,[26] Povinelli writes that "the intimate couple is a key transfer point between, on the one hand, liberal imaginaries of contractual economies, politics, and sociality, and, on the other, liberal forms of power in the contemporary world."[27] For Povinelli, the conjugal couple—for example, husband and wife—is a node connecting liberal beliefs to resource distribution. That is, couples formed through choice receive more benefits from state and economy, while "traditional" relations receive less. This uneven distribution of resources to different populations in turn contradicts the liberal principles of individual freedom or egalitarianism championed by settler societies.

Povinelli argues that liberal societies map global intimacy according to criteria of individual freedom, social constraints, and contract. The achievement of individual choice in intimate love is located in the West in this symbolic map, although in ways that are balanced with respect for social constraints. Excessive constraint from families or tradition is located in non-Western cultures. This map of intimacy recreates the territorial, temporal, and civilizational otherness of the global south.[28]

These approaches to intimacy share characteristics that mark an analytical use of the concept of intimacy. Influenced by queer theory, they investigate the ways that heterosexuality acts as a norm governing social life (that is, heteronormativity). Critical studies of intimacy also consider the ways that race, gender, and nation operate together rather than as separate social arenas, just as they show how social life is divided into public and private domains rather than assume those divisions as given.[29] The feminist and queer lens in these works enriches the historical sociological account of the erosion of the infrastructure for public life by adding attention not only to race, sex, and gender, but also to the ideological and affective dimensions of altered conditions for intimacy.

TRANSNATIONAL INTIMACY

Using the United States or Australia or Europe to discuss global intimacy would be disappointing if it intended these Western powers to stand in for the world—that is, to suggest that it portrays the cutting edge of intimacy that other regions will follow. The descriptions of intimacy presented here do not intend the United States to represent but to illustrate how political economic changes involve intimacy. They therefore understand U.S. social patterns within a transnational frame that includes the aftermath of colonial empires and slave economies and attends to broad shifts associated with the global economy and international relations. These shifts include the globalization of manufacturing from the (partially) unionized global north to the (mostly) nonunionized global south. The functions of the state have been rearranged, with much of state provisioning reassigned to corporations (subcontracted military services) or nonprofits, including religious institutions, and in general reducing the government's role in channeling national resources to mitigate the uneven effects of a capitalist economy or historical inequality. Such transformations changed livelihoods, communities, and allegiances, altering the conditions for and the norms about intimacy felt with fellow citizens but also with kin, neighbors, and friends.

Many scholars, particularly in the United States, prefer the concept of transnational to the concept of global or international.[30] The turn to transnational analysis was sparked empirically by pronounced changes in political economy, social formation, and cultural currents worldwide. The post-1970 period of "globalization" has intensified zones that supersede yet rely on nations—for instance, migration flows, multinational corporations, human rights networks, security apparatus, and the European Union. Conventionally, discussions of sex and gender systems have often used the nation-state as the unit of analysis (often implicitly, as with much U.S-based work that fails to register the specificities of the American location). Transnational analysis considers how national borders are established and reinforced as well as how various flows cross them. The mainstream concept of the international can obscure ways that local life is affected by phenomena that are not confined to the nation state. Scholars who are more skeptical about taking national sovereignty as a given in descriptions of global phenomenon use the term *transnational* to convey an image of the world made up of flows that transverse national boundaries.

Rethinking national and global domains has implications for thinking about intimate realms. The increase in links across countries suggests that intimate life takes place through flows and sites that cross national borders—for example, through sex tourism or the dispersal of family members in different countries. Such rethinking has stimulated reflection on the relations between queer life and the reinforcing or crossing of national borders.[31] Extending these insights back in time, scholars have reconsidered intimate life in the emergence of states, nations, and empires.[32] Intimacy, therefore, offers a basis for rethinking national and transnational phenomena.

Ann Stoler's body of work has helped shape this direction. Her work insists that colonial regimes operated intimately—that is, that they affected sexual desire, child rearing, and family life. The intimate domain was central to colonial rule. European families living in the colonies were anxious about how to raise their children—nurtured by native nannies and playmates—as Europeans. Intimate life was a weak link in the precarious reproduction of European identities and the racial justification of colonial rule.[33] Stoler's historical analysis has influenced other historical studies[34] as well as research on contemporary investments in intimacy. Contemporary studies explore how facets of modernity, neoliberalism, and capitalism shape prevailing conceptions of intimacy and, vice versa, how ideals for intimate life are intricate components of global political economy.

Neville Hoad's book *African Intimacies* explores the problems of applying the concept of sexuality to the non-Western world, particularly in a region subject to the sexualizing and racial gaze of Europe. His point is not only that words describing sexuality, such as "gay," are culturally specific and should not be imposed on other societies, but that sexuality itself as way to categorize people was developed in Europe and had political uses. The use of sexuality to describe African societies, then, cannot be separated from ongoing histories of European imperialism because it involved codifying African sexuality as an object for discussion, evaluation, and regulation.[35] Hoad's analysis embeds modes of relatedness within contexts of global powers, notably colonial and ongoing imperial forces of Europe and the United States, and views knowledge about intimacy as itself part of that broader context.

INTIMATE ECONOMIES

Cultural studies of the politics of intimacy, like those of Berlant, have focused especially on "the rhetorics, laws, ethics and ideologies of the hegemonic public sphere."[36] Others have extended this critical analysis of global intimacy by exploring intimacy in conjunction with the economy, particularly the effects of transnational capitalism around the world.

In hegemonic understandings, by which I mean those emerging through Western-dominated versions of modernity, the intimate and economic are different realms. Common sense understands the intimate realm to be *noneconomic*, or at least nonproductive: part of what defines the values of family, romance, and friendship is their stark difference from market values (a distinction that attributes masculinity to the market sphere and femininity to the domestic sphere). (These intimacies are also often conceived as nonstate and nonpolitical.) To continue with the generalization, this deep logic expects the private nuclear family to buffer the cold marketplace by offering its workers the sustenance of human relations motivated by affection, not instrumentality. And often they do.

Modern economies depend on these divisions, in fact. The mark of modern retail and service industries is the removal of signs of workers' personal life: you should not see where workers take their meals in a modern establishment. As the New Zealand feminist advocate Marilyn Waring has noted, a national economy is measured in terms of productivity according to market exchange (even illegal economies can be counted), while the work and exchanges that people do in intimate, nonmarket contexts is not counted.[37]

The notion that market forms are nonintimate, generic, and potentially universal presents one of the most powerful and effective discourses worldwide. This hegemonic view—enduring even after bubble economies burst dramatically—discourages inquiries into the interaction of capitalism and intimacy. Virtually by definition, capitalism is seen as external and impersonal, at least in the West. Fast-food chains, airport terminals, and shopping malls have become seemingly neutral backdrops in the first world, although in the third world they are marked as an incursion of Western modernity, for better or worse.

Neoclassical economics holds that capitalist economic forms offer a neutral ve-hicle for diverse peoples to realize security, happiness, and democracy; markets realize people's desires without radically altering what orthodox theories of mo-dernity consider to be intimate life: notably heterosexual couples, nuclear fami-lies, and domestic realms that, by definition, are separated from capitalist econ-omies (except through consumption).

The models for economically developing third-world countries known as modernization theory followed much of this logic. To achieve capitalist moder-nity, modernization theory called for separating intimate relations like kinship from (wage) work and governance. One example would be the shift from family farm to agro-business. In retail, it manifests as the evolution from shophouse to department store. The shophouse was a prevalent business form in the nineteenth and twentieth centuries, a two- to three-story building where a family lived, made or processed goods, and sold them—that is, it combined production, distribu-tion, and consumption, economic functions that were ideally to be separated in a "modernizing" economy. Typically it was not home to a nuclear family: rather, multiple generations and hired hands typically lived and worked together. As capitalism "penetrates" local society, it disaggregates practices that traditionally have been comingled into different spheres: state, market, and domestic.[38] Fam-ily owners do not live in a department store. Modern businesses hire wage-labor, which belongs to the formal market economy, ideally separated from the "public" sphere of government. Activities associated with kin and community—activities that had been intertwined with economic operations of the shophouse or family farm—become relegated to the domestic realm of households, a "private" domain whose ideal economic roles are confined to savings and consumption.[39] Major policies for development in the third world thus intended to rearrange intimate functions and locate them in a private, nonpolitical realm. At the same time, Amy Lind's queer critique of the development industry argues that it prioritizes "the heterosexual household as the foundation of family reproduction and survival" and sees this version of intimacy as the key to growing a nation's economy.[40]

Denaturalizing the universality of capitalism, therefore, involves advancing an alternate vision of the nexus of intimacy and the global economy. Contrast-ing approaches to the economy insist that it is not separate from intimate life but is implicated in social relations and identities. These approaches take various forms in feminist theory, queer theory, cultural anthropology, human geography, and heterodox economics, including Marxism. In these frameworks, economic systems are social and cultural processes that affect, and are shaped by, intimate relations.[41]

Classical anthropology emphasized that kinship has economic functions, particularly in societies considered technologically primitive and without a state. More recent ethnographic studies have demonstrated how global capitalism is made from social relationships, including intimate relationships—that is, that the formal, public "economy" depends on and is inextricable from the realm considered, in modern sensibilities, "private." Indeed, modern capitalist markets emerged in and through class, gender, and ethnic relations, relations that are defined and sustained through different forms of intimacy. Studies of women factory workers have shown how kinship and gender are woven into the frontier of industrial modernity in Southeast Asia, for example.[42]

In my ethnography of Bangkok, I proposed the term *intimate economies* as a rubric for analyzing the ways that intimate and economic life, presumed separate, in practice overlap. Intimate life (for example, relations associated with gender, sexuality, or ethnicity) crosses into the "public" sphere of markets and jobs, while those public realms profoundly affect people's "private" interactions and self-conceptions. Numerous works demonstrate how the public sphere of state and markets affects intimate life in locations around the world. The efforts to demonstrate the converse, how intimacy shapes the public sphere (or globalization) have increased. Magdalena Villarreal, for example, argues that social relations are involved in "the processes of negotiation and creation of value" itself. Studying rural Mexico, she sees "social resources as currencies which are situationally attributed exchange value."[43] For example, the interpretation of a customer's relations with the shopkeeper and others in the community determines whether or not he receives credit or what he pays for goods. Through this Latin American example, Villarreal inverts the model of market exchange penetrating economies of affection by demonstrating how social life produces literal market values.

An ethnographic understanding of the interaction between intimate life and the global economy also requires recognizing that there is not just one economic form but a plurality of economic modes that inform local life. "Capitalism's others," as J. K. Gibson-Graham put it, include the economies of affection, also known as kin economy, folk economy, or moral economy.[44] Much of canonical cultural anthropology has been dedicated to elaborating the patterns, categories, and roles that define systems of exchange other than the capitalist market system. Classical studies on kinship, for example, chronicle the rules and regulations governing the exchange of wealth, resources, and women as wives among groups. Indeed, the exchange of women has been seen as foundational to culture and society itself.[45] In Thailand, the practice and ideal of exchange organizes a number of relationships: parent-child (including children's, especially daughters', indebtedness to parents), senior-junior, husband-wife (and her family), laity-monkhood, human-spirit world, and friendship. Different economic systems structure intimacy differently. Moral economies and economies of affection are not (generally) guided by extracting profit, technically speaking, although families may accumulate great wealth. Rather, their aim is to reproduce intimacy with the human and spiritual world.[46]

Capitalism differs from its economic others precisely in how it involves intimate relations. A common observation of third-world economies or "traditional societies" is that their markets and states are deeply intertwined with kinship and ethnicity, an arrangement that has been called an "economy of affection." Even with capitalist development, economies of affection continue to exist. Modernization theories read these as (problematic) cultural survivals that are obstacles to the separation of public and private realms—that is, modernization theory evaluates societies in relation to how much intimate life is kept apart from economic or government functions. Other interpretations see such ad-mixed systems as typical of colonial and postcolonial cultures and not at all out of keeping with capitalism as usual.[47] Take the phenomenon labeled "crony capitalism" in third-world societies. Nepotistic business practice presents an example of economies of affection changing and thriving within capitalist development.[48]

As only one mode of organizing provisioning or material transactions (albeit an aggressive one), capitalism necessarily interacts with other systems of organizing work and exchange. Modernization theory sees the mix of intimate life with nonmarket economies as a backward economic form that needs to be supplanted by capitalist markets in order to progress.[49]

Many progressive critics of capitalist development value economies of affection positively. Nonmarket, nonstate versions of public intimacy can be the basis for challenging exploitation. When capitalist development imposes more impersonal relations in public life, people assert the legitimacy of their precapitalist modes of public intimacies. The term *moral economy* was coined to convey the ways that noncapitalist expectations about public sociality have been a source of resistance to capitalist developments.[50] In a similar vein, Aihwa Ong interprets the outbursts of spirit possessions in Southeast Asian factories as an embodied critique of the dehumanized relations entailed in industrial manufacturing.[51]

The attempt to reconstitute "traditional" intimate life can lay claim to a return to authentic origins. The ways that traditions for intimacy are interpreted or lived out are by-products of modern life. As such, they often reflect hierarchical evaluations of good and bad modes of intimacy. Thai heterosexual relations have transformed markedly in relation to twentieth-century political economic forces. Migration from China to Siam-Thailand changed from mainly single male sojourners to married couples; the bride gift from a groom to the bride and family underwent enormous inflation; courtship became more erotic and oriented to dating couples; and the suburban nuclear family has become an emblem of modern citizenship.[52] The heterosexual couple has come to prominence in recent years, in social space granted to male-female dating and in representations of families that center more on the husband-wife pair. Despite the modernity of this version of heterosexuality, this form of intimacy has been rendered as authentic and traditional, while the apparent modernity of homosexual intimacy—as *gay* or *tom* or *dee* identities—renders them less authentic, less "Thai," and linked with materialist values associated with capitalist modernity: selfishness and a lack of control.[53] The *kathoey*—a term that commonly refers to male-to-female transgender individuals or transsexuals (or to transgenderism in general)—is a figure of long-standing history in Thai society: yet the *kathoey* has never become emblematic of Thai tradition in official discourse. What counts as traditional forms of intimacy, in other words, is both selective and defined in and by the present.

The heterosexualization of society presents a typical pattern. Modern versions of heterosexuality, indelibly shaped by Western modernity (for example, in the white wedding format or the concept of heterosexuality as psychologically normal) are embraced as manifestations of tradition in many societies, while homosexual intimacies, even when forms of same-sex sexuality had been integrated in historical practice, are rejected as foreign. One example is the state homophobia found in Africa in Zimbabwe, Uganda, and other countries, which assert tradition but are the product of colonial rule, market disruption, and globalizing, U.S.-politicized religious campaigns.[54] The point is that celebrations of traditional ways of life as sources of resistance to global capitalism often overlook the ways that moral economies are not only moral but moralistic and exclusionary as well.

The study of global intimacies has methodological dimensions. To study global intimacies, where do you look? Many studies of the effects of globalization on intimate spheres look to family and village life, sites understood to be penetrated by global capitalism, while others focus on expressions of same-sex sexuality or transgender identity. Another ethnographic approach targets the sites of capitalist modernity itself, arguing that the interplay of global markets and intimate life should be studied *within* public sites themselves. As Berlant says, "Institutions not usually associated with feeling can be read as institutions of intimacy."[55] Applying Berlant to the global economy, the banal, deliberately generic, modern sites of corporate capitalism—fast-food franchises, airports, stock market trading floors, or shopping malls—might be recognized not only as signs of eroding intimacy, but as significant stages for intimate life as it plays out in the present. Indeed, by now, more intimate life transpires in Thailand's shopping malls than in its Buddhist temples.[56] Thick descriptions of quotidian public culture illuminate the intimate dimensions of global modernity.

INTIMATE REPLACEMENTS

The term *intimacy* commonly provides a synonym for a concept of proximate, close relations, often connected with the interior and the personal: sexual and romantic relations; "local," microlevel, or proximate relations; "private life"; embodied life;[57] or psychological dimensions. The critical use of intimacy as an analytical term draws on this set of meanings but in ways that critique prevailing conceptions of globalization, governments, and personal life. In so doing, these studies attempt to provide conceptual depictions of the intimate nature of states, economies, and globalization. In her germinal 1988 essay, "Gender: A Useful Category of Analysis," Joan Scott argues that gender is useful as "a primary way of signifying relationships of power."[58] The emergence of *intimacy* as a term common to critical scholarship suggests that it offers another way to signify relations of power, a way that subsumes, or differs from, available critical concepts like gender or sexuality.

The proliferation of the category of intimacy reflects dissatisfaction with inherited terms and the theories behind them. Therefore, one way to understand what intimacy is doing in transnational analysis is to identify what intimacy is *not*.[59] The use of intimacy aims to avoid replicating problems identified in mainstream and radical depictions of social life. Such concepts include identity, kinship, and the public/private divide. The turn to intimacy speaks to scholars' desire for a flexible term that allows new descriptions that do not reify nation, identity, family, or related categorical units. In this way, identity is a placeholder defined as much in negative terms—what the author does not want her description to do—as by any delimited content.

In prevalent analyses of modernization, intimacy offers a convenient term for demarcating familiar distinctions between public/private and local/global.

"Intimate," in these meanings, contrasts the authenticity of local life with external impositions, usually understood as modern, Western, or capitalist. However critical the intent of these discussions, their writing presupposes that intimacy belongs to the local level or the private sphere. Given that the domestic and private realm is glossed as feminine in the culture of Western modernity, these uses of *intimate* have gendered connotations that often remain unexamined in accounts of family or community life in a context of globalization.

The critical scholarship I have described counters the understanding of intimate as a private, local realm that until recently was cordoned off from broader forces. It does not position intimacy as a private realm in the conventional sense of "private." Instead, this vein of critical global analysis deconstructs the separation of "the economy" from family, home, or private life. Some of the work focuses on public intimacy: "modes of attachment that make persons public and collective and that make collective scenes intimate spaces."[60] Feminist scholars examining the public/private divide have shown how this divide, and the gendered associations with each side, resulted from economic, political, and intellectual influences.[61] Capitalist modernity, including its theories, produced the commonsense view that the private, domestic, and feminine sphere is nonproductive and economically irrelevant. At the same time, this separation constituted modernity. As Berlant puts it, "Liberal society was founded on the migration of intimacy expectations between the public and the domestic."[62] The analysis of intimate economies, for another example, denaturalizes market economies by showing they intermix with intimate life.

For critical scholarship, *intimacy* facilitates analyses that incorporate reflections not only about mainstream concepts but also about terms that emerged through radical theory. In feminist and queer studies, the use of intimacy replaces the rubric of identity and the mandates of identity politics. As Povinelli insists of her work, "This book is not interested in the study of identities so much as it is interested in the social matrix out of which these identities and their divisions emerge."[63] Intimacy emphasizes relationality. Povinelli and other feminist and queer scholars employ intimacy to discuss identity in terms of relationship, presenting gender, race, and sexuality as inextricable from, and realized through, relays of power. They do so in ways that incorporate criticisms of Marxist interpretations that reduced social relations to economics; studies of sociality continue to grapple with ways to recognize the force of political economic contexts without that economic reductionism. More obviously, the turn to intimacy in its erotic meanings results from the way that sexuality, as a domain of knowledge, has been problematized. Specifically, Michel Foucault argued that the modern category of sexuality—as a real phenomenon to be measured, known, and felt—rather than liberating people has been a significant way to subject them to modern social power. Clearly, this critical use of intimacy does not mean scholars eschew gender or sexuality as categories of analysis that illuminate relations of power, in the words of Joan Scott. The examples above depend on the analysis of queer politics, heteronormative mandates, women's bodies, sex/gender regimes, and the "white patriarchal nuclear family," in Wiegman's terms.[64] In particular, scholarship on globalization (such as that found in this volume) continues to demonstrate the raw and subtle ways that gender is involved in transnational processes, a commitment manifest in my own research. But the increasing use of intimacy flags shared desires for rubrics encompassing or sidestepping the specific meanings of gender and sexuality. *Intimacy*, as an unfixed but legible term, works to cover an open-ended array of relations (rather than assuming the couple or family);[65] to avoid assigning identities based on relationship (for example, gay identity based on same-sex practices); and to investigate relationships alongside their categorizations (for example, both experienced family relations and the evaluations of proper kinship).

For anthropology, intimacy joins other efforts to escape the weighty associations that adhere in concepts of kinship. The ethnographic use of intimacy resonates with the concept of relatedness parlayed by new kinship studies.[66] The use of relatedness "in opposition to, or alongside, 'kinship'" flags an intent to discover, rather than assume, which modes of relatedness given peoples find salient and to displace the biological/social binary of kinship concepts.[67]

That intimacy is *not* these particular received concepts, but is also not some other clearly delineated referent, suggests that *intimate* is serving as a placeholder in critical analyses of global life. For now, intimacy allows analysts to look at relational life—including the feelings and acts that comprise it, in relation to political and economic regimes—in conventional sociological terms and to consider both micro and macro levels, although of course the critical study of intimacy eschews this neat division. The term *intimacy* is intended to resist ideological reifications of family, sexuality, or community—that is, to avoid recreating forms of knowledge that perpetuate global inequality. As a placeholder, intimacy allows critical accounts of colonial empire or capitalist modernity because it is a flexible, provisional reference that emphasizes linkages across what are understood to be distinct realms, scales, or bodies. Whether an analytical concept or a placeholder, the critical study of intimacy provides a useful category in the transnational analysis of power.

NOTES

1. Twentieth-century versions of state socialism, such as Maoist regimes, also attempted to choreograph intimacy in order to generate new socialist subjects. Socialist intimacy may be read as a form of modernity, and there is a growing body of fascinating work on intimate relations under command economies of China, the Soviet Union, Cuba, and other socialist countries. However, my emphasis in this exploration of the global political economy of intimacy remains on global capitalism and, for the most part, states organized by liberal political traditions.

2. Ara Wilson, *The Intimate Economies of Bangkok*. Although this essay makes use of field research on Thailand that has spanned more than two decades, it is concerned with analytical more than properly ethnographic questions.

3. Raymond Williams, *Marxism and Literature*.

4. Ann Cvetkovich and Ann Pellegrini, eds., *Public Sentiments. S&F Online* 2, no. 1 (2003). http://www.barnard.columbia.edu/sfonline/ps/intro.htm (accessed 11 November 2009).

5. Hague *v.* CIO, 307 U.S. 496, 515 (1939). In the case, when labor organizers were denied a permit for a public discussion of the National Labor Relations Act, they challenged the Jersey City ordinance requiring permits for public meetings.

6. Haya El Nasser, "Gated Communities More Popular, and Not Just for the Rich," *USA TODAY*, 15 December 2002. *usatoday.com* (accessed 20 August 2010). A partial list of names for gated communities in one county in Florida includes: Harbour [*sic*] Lake Estates; Sunset Lakes; Monarch Lakes; Silver Isles; Silver Falls; Somerset; Windsor Palms.

7. Sven Bislev, "Privatization of Security as a Governance Problem: Gated Communities in the San Diego Region," *Alternatives* 29, no. 5 (2004): 599–618; Edward James Blakely and Mary Gail Snyder, *Fortress America*.

8. Setha M. Low, *Behind the Gates*.

9. David J. Kennedy, "Residential Associations as State Actors: Regulating the Impact of Gated Communities on Nonmembers," *The Yale Law Journal* 105, no. 3 (December 1995): 763.

10. Kennedy, "Residential Associations."

11. For the way that private gated communities are subsidized by the public weal, see Andrew Stark, "America, the Gated?" *WQ* (*The Wilson Quarterly*) 22, no. 1 (1998): 58–79.

12. Kennedy "Residential Associations," 778.

13. Kennedy, "Residential Associations."

14. Easton Town Center, "About Us," http://eastontowncenter.com/AboutUs.aspx (accessed 20 August 2010).

15. Ibid.

16. In 1976 the Supreme Court found that there is no federal constitutional right to free speech in shopping malls. Hudgens *v.* NLRB, 424 U.S. 507 (1976).

17. Robert D. Putnam, *Bowling Alone*.

18. Robyn Wiegman, "Intimate Publics," 860.

19. Ibid., 863.

20. Ibid., 861.

21. Lauren Gail Berlant, *The Queen of America Goes to Washington City*.

22. Lauren Berlant, "Intimacy: A Special Issue," 282.

23. Berlant, "Intimacy," 283.

24. Berlant, *The Queen of America Goes to Washington City*, 210.

25. Povinelli recasts the tension expressed in political theory between individual freedom, which she terms the "autological," and society's constraints, expressed, e.g., in kinship or tradition, which she calls the "genealogical." Her book deconstructs the differential evaluations of these modes of intimacy, identifying "which forms of intimate dependency count as freedom and which count as undue social constraint; which forms of intimacy involve moral judgment rather than mere choice; and which forms of intimate sociality distribute life and material goods and evoke moral certainty." Elizabeth A. Povinelli, *The Empire of Love*, 3.

26. Michel Foucault, *The History of Sexuality*, 103.

27. Memorably, Povinelli writes: "If you want to locate the hegemonic home of liberal logics and aspirations, look to love in settler societies." Povinelli, *The Empire of Love*, 17.

28. Povinelli's book is a sustained critique of Anthony Giddens's writing about intimacy: Anthony Giddens, *The Transformation of Intimacy*. Giddens proposes that intimacy based on choice, romantic love, and egalitarian relations emerged within European modernity. It represents a kind of progress of human freedom and also a check on the instrumental values of the market. For Giddens, the development of romantic conjugal love as the organizing mode for private life is a democratization of human relations. Povinelli's work criticizes Giddens's Eurocentrism, his progress narrative, and his optimistic take on modernity and liberal democracy. As a transnational analysis, her approach insists on placing the intimacy characteristic of liberal democracies in a frame that includes the effects of colonialism, racism, and uneven distribution of resources or value across populations.

29. See, e.g., Wiegman, "Intimate Publics."

30. Inderpal Grewal and Caren Kaplan, eds., *Scattered Hegemonies*; Jasbir K. Puar, *Terrorist Assemblages*.

31. Examples that reflect on the relation of the queer life to national or global scales include: Tom Boellstorff, *The Gay Archipelago*; Martin F. Manalansan, *Global Divas*; Puar, *Terrorist Assemblages*; Gayatri Reddy, *With Respect to Sex*.

32. See, e.g., Foucault, *The History of Sexuality*; Anne McClintock, *Imperial Leather*; Rayna Rapp Reiter, "Gender and Class: An Archaeology of Knowledge Concerning the Origins of the State," *Dialectical Anthropology* 2, no. 4 (1977): 309–16; Victoria Wohl, *Intimate Commerce*.

33. Ann Laura Stoler, "Carnal Knowledge and Imperial Power"; Ann Laura Stoler, ed., *Haunted by Empire*; Ann Laura Stoler, *Race and the Education of Desire*.

34. Examples of critical historical work on intimacy include: Nayan Shah, "Adjudicating Intimacies on U.S. Frontiers." Lisa Lowe has deconstructed European and North American notions of intimacy, arguing that colonialism and slavery produced intimacies

among the subjugated that, while denigrated, underwrote the possibilities for a liberal European understanding of intimacy: Lisa Lowe, "The Intimacies of Four Continents."

35. Neville Hoad, *African Intimacies*.

36. Berlant, "Intimacy," 283.

37. Marilyn Waring, *If Women Counted*.

38. Linda Nicholson offers a model for analyzing public/private formation under capitalist economies: Linda Nicholson, "Feminism and Marx." See also Karl Polanyi, *The Great Transformation*.

39. See, e.g., Wilson, *Intimate Economies*.

40. Amy Lind, "Governing Intimacy, Struggling for Sexual Rights." Also see Kate Bedford's discussion of ways that World Bank policies concerning the family target racialized and poor men as problems for realizing family functions that will promote economic growth: Kate Bedford, *Developing Partnerships*.

41. Canonical anthropology defines the "economy" as systems of production, distribution, and consumption, a definition that recognizes that societies have been arranged by different economic principles, not all of them governed by a market logic, such as the ways kinship shapes the gendered division of labor, and such as work and exchange associated with a feminized domestic sphere and not officially tallied as in standard economic measurements. Feminist economists have defined the economy as the system for provisioning society: Drucilla K. Barker and Susan Feiner, *Liberating Economics*.

42. Aihwa Ong, *Spirits of Resistance and Capitalist Discipline*; Mary Beth Mills, *Thai Women in the Global Labor Force*; Carla Freeman, *High Tech and High Heels in the Global Economy*; Lisa Rofel, "Qualities of Desire."

43. Magdalena Villareal, "Cashing Identities in the Non-material World of Money."

44. E. P. Thompson, *Customs in Common*.

45. Paul Bohannan, "The Impact of Money on an African Subsistence Economy"; Karl Polanyi, *The Great Transformation*; Gayle Rubin, "The Traffic in Women."

46. The idiom and mechanism for many, if not most, relationships in Thailand is exchange. Prominently, the relationship between the monastic order and the laity is depicted as one of exchange. Householders (mainly women) provide the daily sustenance to monks, who act as "fields of merit," providing the opportunity to accumulate merit (which is calculated quite materially in terms of a store or amount of substance). The enactment and definition of many Thai social identities, such as a woman's position as "nurturer" or the relations of seniors to juniors, can also be understood as an orientation framed in terms of debt and exchange. Anyone can incur debt to a guardian or one who has offered significant aid and instruction (e.g., teachers), and all children are born indebted, but male and female children have different prospects for repaying that debt.

47. Dipesh Chakrabarty, *Provincializing Europe*.

48. Nepotism reflects the place of close ethnic and family networks in business, while graft represents locally calibrated expectations of exchange. But here the model of economic interaction is illuminating: crony capitalism and routine graft are generated through local codes interacting with, and being distorted by, capitalism, specifically by global finance.

Pasuk Phongpaichit and Chris Baker, *Thailand: Economy and Politics*. (It is also worth remembering the corrupt operations of heralded companies within the United States, memorably Enron, Worldcom, and Bernie Madoff's enterprise, among other infamous examples.)

49. One example of modernization theory's view of local economies as constraints on development is Göran Hydén, *No Shortcuts to Progress*.

50. On the moral economy, see E. P. Thompson's landmark work—e.g., the discussion of the term in *Customs in Common*. Also see James C. Scott, *Weapons of the Weak*. For the ways capitalism changes the social codes of the market in Europe, see Polanyi, *Great Transformation*. The "moral economy" does not mean a "natural economy" that is unchanging and apart from a market economy.

51. Ong, *Spirits of Resistance*.

52. See Wilson, *Intimate Economies*, chapter 3.

53. See Megan Sinnott, *Toms and Dees*.

54. Hoad, *African Intimacies*; Jeff Sharlet, "White Man's Burden."

55. Berlant, "Intimacy," 283.

56. Especially the intimate life of women, who are excluded from being ordained as monks and allowed only an inferior, spiritually ineffective role glossed as nuns. Given that women have long predominated in the vernacular markets in Thailand, the market sphere is pivotal to the female gender and women's intimate lives. See Wilson, *Intimate Economies*.

57. Andrea Whittaker's able ethnographic study is an example of this use of intimacy as associated with feminine bodies in particular: Andrea Whittaker, *Intimate Knowledge*.

58. Joan Scott, "Gender: A Useful Category of Historical Analysis."

59. In my discussion of intimacy as an analytical category, I am indebted in particular to feminist discussions of investments in particular terms or knowledge objects, such as those of Clare Hemmings and Robyn Wiegman. See, e.g., Clare Hemmings, "Invoking Affect"; Wiegman, "Intimate Publics." For an overview of feminist understandings of global intimacy, see G. Pratt and V. Rosner, "Introduction: The Global and the Intimate" and Geraldine Pratt and Victoria Rosner, in the introduction to this volume.

60. Berlant, "Intimacy."

61. On women's rights and the public/private divide, see Donna Sullivan, "The Public/Private Distinction in International Human Rights Law."

62. Berlant "Intimacy," 284.

63. Povinelli, *The Empire of Love*, 4.

64. Wiegman, "Intimate Publics."

65. E.g., scholars in the United Kingdom in particular have turned to the rubric of citizenship to discuss intimacy. In his book *Telling Sexual Stories*, Ken Plummer uses the term "intimate citizenship" in a discussion centered on the increasing power of narratives in defining sexual experience and sexual politics in the United Kingdom: Ken Plummer, *Telling Sexual Stories*. See also Jeffrey Weeks, "The Sexual Citizen"; Phil Hubbard, "Sex Zones"; David Bell and Jon Binnie, *The Sexual Citizen*. These works emphasize the het-

eronormativity of civic life and also the struggles of queer, gay, and lesbian communities
to claim public space and state recognition.

66. Janet Carsten, *Cultures of Relatedness*; *After Kinship*.
67. Carsten, *Cultures of Relatedness*.

BIBLIOGRAPHY

Barker, Drucilla K., and Susan Feiner. *Liberating Economics: Feminist Perspectives on Families,
Work, and Globalization*. Ann Arbor: University of Michigan Press, 2004.

Bedford, Kate. *Developing Partnerships: Gender, Sexuality, and the Reformed World Bank*.
Minneapolis: University of Minnesota Press, 2009.

Bell, David, and Jon Binnie. *The Sexual Citizen: Queer Politics and Beyond*. Cambridge: Polity
Press, 2000.

Berlant, Lauren. "Intimacy: A Special Issue." *Critical Inquiry* 24, no. 2 (Winter 1998): 281–88.

——. *The Queen of America Goes to Washington City: Essays on Sex and Citizenship*. Series Q.
Durham: Duke University Press, 1997.

Bernstein, Elizabeth. *Temporarily Yours: Intimacy, Authenticity, and the Commerce of Sex*.
Chicago: University of Chicago Press, 2007.

Blakely, Edward James, and Mary Gail Snyder. *Fortress America: Gated Communities in the
United States*. Washington DC: Brookings Institution Press, 1997.

Boellstorff, Tom. *The Gay Archipelago: Sexuality and Nation in Indonesia*. Princeton: Princeton University Press, 2005.

Bohannan, Paul. "The Impact of Money on an African Subsistence Economy." *Journal of
Economic History* 19 (1959): 491–503.

Carsten, Janet. *After Kinship: New Departures in Anthropology*. Cambridge: Cambridge University Press, 2004.

——. *Cultures of Relatedness: New Approaches to the Study of Kinship*. Cambridge: Cambridge
University Press, 2000.

Chakrabarty, Dipesh. *Provincializing Europe: Postcolonial Thought and Historical Difference*.
Princeton: Princeton University Press, 2000.

Cohen, Lawrence. "Supplementarity: Hospitable Care in the Apollonian Age." *Body & Society*,
forthcoming.

Foucault, Michel. *The History of Sexuality*. 1st American ed. New York: Pantheon Books,
1978.

Freeman, Carla. *High Tech and High Heels in the Global Economy: Women, Work, and Pink-
Collar Identities in the Caribbean*. Durham: Duke University Press, 2000.

Giddens, Anthony. *The Transformation of Intimacy: Sexuality, Love, and Eroticism in Modern
Societies*. Stanford: Stanford University Press, 1992.

Grewal, Inderpal, and Caren Kaplan, eds. *Scattered Hegemonies: Postmodernity and Transnational Feminist Practices*. Minneapolis: University of Minnesota Press, 1994.

Hemmings, Clare. "Invoking Affect: Cultural Theory and the Ontological Turn." *Cultural Studies* 19, no. 5 (2005): 548–67.

Hoad, Neville. *African Intimacies: Race, Homosexuality, and Globalization.* Minneapolis: University of Minnesota Press, 2007.

Hubbard, Phil. "Sex Zones: Intimacy, Citizenship, and Public Space." *Sexualities* 4, no. 1 (February 2001): 51–71.

Hydén, Göran. *No Shortcuts to Progress: African Development Management in Perspective.* Berkeley: University of California Press, 1983.

Lind, Amy. "Governing Intimacy, Struggling for Sexual Rights: Challenging Heteronormativity in the Global Development Industry." *Development* 52, no. 1 (March 2009): 34–42.

Low, Setha M. *Behind the Gates: Life, Security, and the Pursuit of Happiness in Fortress America.* New York: Routledge, 2003.

Lowe, Lisa. "The Intimacies of Four Continents." In Ann Laura Stoler, ed., *Haunted by Empire: Geographies of Intimacy in North American History,* 191–212. Durham: Duke University Press, 2006.

Manalansan, Martin F. *Global Divas: Filipino Gay Men in the Diaspora.* Perverse Modernities. Durham: Duke University Press, 2003.

Massey, Doreen B. *Space, Place, and Gender.* Minneapolis: University of Minnesota Press, 1994.

McClintock, Anne. *Imperial Leather: Race, Gender, and Sexuality in the Colonial Contest.* New York: Routledge, 1995.

Mills, Mary Beth. *Thai Women in the Global Labor Force: Consuming Desires, Contested Selves.* New Brunswick: Rutgers University Press, 1999.

Nicholson, Linda. "Feminism and Marx: Integrating Kinship with the Economic." In Seyla Benhabib and Drucilla Cornell, eds., *Feminism as Critique: On the Politics of Gender,* 16–30. Minneapolis: University of Minnesota Press, 1987.

Ong, Aihwa. *Spirits of Resistance and Capitalist Discipline: Factory Women in Malaysia.* Albany: State University of New York Press, 1987.

Padilla, Mark. *Love and Globalization: Transformations of Intimacy in the Contemporary World.* Nashville: Vanderbilt University Press, 2007.

Peters, Julie, and Andrea Wolper, eds. *Women's Rights/Human Rights: International Feminist Perspectives.* New York: Routledge, 1995.

Phongpaichit, Pasuk, and Chris Baker. *Thailand: Economy and Politics.* Oxford: Oxford University Press, 1995.

Plummer, Ken. *Telling Sexual Stories: Power, Change, and Social Worlds.* London: Routledge, 1995.

Polanyi, Karl. *The Great Transformation: The Political and Economic Origins of Our Time.* New York: Beacon Press, 1944/1957.

Povinelli, Elizabeth A. *The Empire of Love: Toward a Theory of Intimacy, Genealogy, and Carnality.* Durham: Duke University Press, 2006.

Pratt, G., and V. Rosner. "Introduction: The Global and the Intimate." *WSQ* 34 (2006): 13–24.

Puar, Jasbir K. *Terrorist Assemblages: Homonationalism in Queer Times*. Durham: Duke University Press, 2007.

Putnam, Robert D. *Bowling Alone: The Collapse and Revival of American Community*. New York: Simon and Schuster, 2000.

Reddy, Gayatri. *With Respect to Sex: Negotiating Hijra Identity in South India*. Worlds of Desire. Chicago: University of Chicago Press, 2005.

Rofel, Lisa. "Qualities of Desire: Imagining Gay Identities in China." *GLQ* 5, no. 4 (1999): 451–74.

Rubin, Gayle. "The Traffic in Women: Notes on the Political Economy of Sex." In Rayna R. Reiter, ed., *Toward an Anthropology of Women*, 157–210. New York: Monthly Review Press, 1975.

Scott, James C. *Weapons of the Weak: Everyday Forms of Peasant Resistance*. New Haven: Yale University Press, 1985.

Scott, Joan. "Gender: A Useful Category of Historical Analysis." *American Historical Review* 91, no. 5 (1986): 1053–75.

Shah, Nayan. "Adjudicating Intimacies on U.S. Frontiers." In Ann Laura Stoler, ed., *Haunted by Empire: Geographies of Intimacy in North American History*, 116–39. Durham: Duke University Press, 2006.

Sharlet, Jeff. "White Man's Burden." *Harper's Magazine*, September 2010.

Sinnott, Megan. *Toms and Dees: Transgender Identity and Female Same-Sex Relationships in Thailand*. Honolulu: University of Hawaii Press, 2004.

Stoler, Ann Laura. "Carnal Knowledge and Imperial Power: Gender, Race and Morality in Colonial Asia." In Micaela diLeonardo, ed., *Gender at the Crossroads of Knowledge*, 51–101. Berkeley: University of California Press, 1992.

——, ed. *Haunted by Empire: Geographies of Intimacy in North American History*. American Encounters/Global Interactions. Durham: Duke University Press, 2006.

——. *Race and the Education of Desire: Foucault's History of Sexuality and the Colonial Order of Things*. Durham: Duke University Press, 1995.

Sullivan, Donna. "The Public/Private Distinction in International Human Rights Law." In Julie Peters and Andrea Wolper, eds., *Women's Rights, Human Rights: International Feminist Perspectives*, 126–34. New York: Routledge, 1995.

Thompson, E. P. *Customs in Common*. New York: New Press, 1993.

Villareal, Magdalena. "Cashing Identities in the Non-material World of Money." English-language manuscript in author's possession, no date, p. 8. Published in Danish: Magdalena Villarreal, "Brug af identiterer i den ikke materielle pengeverden" en *Fattigdomsbekaempelse: Nar de fattige udfordrer, Den Ny Verden* 2, August 2001, Copenhagen, Center for Udviklingsforskining.

Waring, Marilyn. *If Women Counted: A New Feminist Economics*. San Francisco: Harper and Row, 1988.

Weeks, Jeffrey. "The Sexual Citizen." *Theory, Culture & Society* 15, no. 3 (August 1998): 35–52.

Whittaker, Andrea. *Intimate Knowledge: Women and Their Health in North-East Thailand*. St. Leonards: Allen and Unwin, 2000.

Wiegman, Robyn. "Intimate Publics: Race, Property, and Personhood." *American Literature* 74, no. 4 (December 2002): 859–85.

Williams, Raymond. *Marxism and Literature.* Marxist Introductions. Oxford: Oxford University Press, 1977.

Wilson, Ara. *The Intimate Economies of Bangkok: Tomboys, Tycoons, and Avon Ladies in the Global City.* Berkeley: University of California Press, 2004.

Wohl, Victoria. *Intimate Commerce: Exchange, Gender, and Subjectivity in Greek Tragedy.* Austin: University of Texas Press, 1997.

2

IN THE INTERESTS OF TASTE AND PLACE
ECONOMIES OF ATTACHMENT

Elspeth Probyn

EATING INTO THE GLOBAL

There is perhaps no area where the global inserts itself more into the intimate lives of people than in the realms of food production and consumption. While the worldwide circulation of food and people has been speeded up in a spectacular fashion through global technology and logistics, food has long traveled and connected very different peoples and lands. Whether it be in the Roman Empire, which circulated different crops and foodstuffs, or in the spice trade that brought Asia, India, and Europe into close contact, or the slave routes of sugar and other alimentary commodities, foodways have long been a privileged way for the proximities of place and taste to intertwine. As the economic historian Harold Innis argued in the 1920s and 1930s, staples such as fish were integral to the development of the economic world as we know it. Innis's classic thesis on the cod fisheries of the Canadian Maritimes demonstrated the far-reaching effects of one staples economy.[1] The Grand Banks of Newfoundland became a battleground between the Old World powers: "Cod from Newfoundland was the lever by which she [England] wrested her share of the riches of the New World from Spain."[2] In Innis's terms, food staples caused empires to rise and fall as their routes rearranged the political and economic face of the world.

The role of food has continued to be the site through which global politics are played out in local ways. The field of food, understood in Pierre Bourdieu's sense as a field of forces and a field of struggles can be seen as a minefield, which pits gender, globalization, and class privilege.[3] In the past thirty years, enhanced technology and political intervention, especially in the United States, have resulted in amazingly cheap and plentiful food for the West. Of course, this has produced ecological catastrophes around the globe as the turn to monocrops and industrialized agriculture continues to destroy the world's cultural and environmental biodiversity. And the Westernization of diets around the world has arguably homogenized tastes and produced a putatively global epidemic of obesity, with effects felt particularly by the poor and often by women.[4]

Conversely "feel-good" food politics have increased in many sections of the developed world, with the rise of farmers' markets and other "alternative" food outlets in most urban Western regions. This plays out in public debates and at times in polemics about the "food miles" and carbon footprints of what we eat. As I've argued elsewhere, this has a tendency to become a fetishization of the local, which can lead to a rather parochial politics of food.[5] For instance, the deeply intertwined economies and livelihoods of small farmers in developing nations often depend on the supply chains of specialty foodstuffs for the West. Their produce is now being condemned because of a simplistic view of "food miles." This means that, for instance, Kenyan farmers of organic baby green beans are being threatened by supposedly progressive organizations such as the U.K.'s Soil Association, which may take away the prized organic sticker solely on the basis of the carbon costs of transport. Little does it matter that the average carbon footprint of someone in Kenya or elsewhere in the developing world is minuscule compared to that of the average Briton. This is what Michael Winter refers to as a "defensive localism," which he sees as emerging from a mixture of "parochialism and nationalism."[6] Class, gender, and ethnic privileges and distinctions are often overlooked in this putative progressive food politics. One of the original founders of the Italian Slow Food movement, Fabio Parasecoli, warns that the place of women in slow food is worrisome: "In a world doomed by pollution, biological homogenization and globalization, women are transformed into the defenders of the holy environment that constitutes the family." Their role would be to stay at home and protect it against the evil forces that haunt our present.[7]

Julie Guthman's critique of the neoliberal force of such politics has incisively revealed the moralistic tone and the ethnicity- and class-blindness of many in the "alternative" food circles.[8] Writing about the rise of the organic movement in northern California, which is most successful in terms of organic salad mix or what the growers call "yuppie chow," Guthman questions the motivation of "those whose moral sensibilities increasingly privileged environmental concerns over social ones."[9] The key thrust of her critique concerns the ways in which the privileged have taken on the congratulatory mantle of self-reflexivity in eating at the expense of those others who are said to lack the capacity to know that local, organic, slow food *tastes* better. With her colleague Melanie Dupuis, Guthman performs a magisterial critique of the politics of fat that links the questions of availability and access to "better food" though the vectors of gender and class that so overdetermine food politics now.[10] Likewise, Lauren Berlant's recent work on the management of obesity skewers the "cruel optimism" where 'living increasingly becomes the scene of the administration, discipline, and recalibration of what constitutes health' for the vast majority of ordinary people.[11] What Berlant calls "slow death" is fed by "particular modes of sweetness and fat [that] are metabolized with particular inefficiency and toxic effect by the human body, and, since they produce more fat storage and cravings for more both in terms of mouthfeel and in terms of insulin instability, the phrase *supply and demand* could easily be replaced with *supply and need*."[12]

The importance of such feminist critiques is undeniable, especially given the ways in which food consumption and production adversely affect women and the poor. Their aim is understandably to scrutinize the subjectivities of women under circumstances not of their own making. Given the vastness and the intricacy of the pressures faced by so many, that other mainstay of feminist argument—agency—becomes ever harder to locate. In the end many analyses return to previous politics, and as Guthman reminds us, the question of food is still very much an institutional and structural problem exacerbated to an impossibly high degree by the global flow of the supply chain, which will not be solved by looking the farmer in the eye at your local market.[13] In Berlant's argument, agency moves sideways and is fleetingly found in moments of "episodic refreshment, for example, in sex, or in spacing out, or in food that is not for thought."[14] Obviously, this is not a particularly satisfying way to live, and as a political response her conclusion comes close to some previous feminist arguments, which privileged feint and irony as feminist and feminine modes of resistance.

Like many feminists, much of my own work has been on questions of subjectivity and was prompted by Adrienne Rich's call to "begin . . . with the geography closest-in."[15] I responded to Rich's politics of location and the politics of positionality with detailed descriptions of my body. That this body, my body, had a tangential relationship with eating led me into analyses of anorexia and control ("The Anorexic Body," 1987) and then away from food and into sex (*Outside Belongings*, 1996) and then back to food *and* sex (*Carnal Appetites: FoodSexIdentities*, 2000). Working through a Foucauldian and then a Deleuzian frame, I sought to examine subjectivity obliquely through eating.

However, in this chapter I want to go beyond my previous work, and also beyond a prevalent feminist focus on questions of human sovereignty and agency under the regime of neoliberalism. With Gerry Pratt and Victoria Rosner, I want to go "beyond the usual register of map-reading" (at which I am atrocious) to try to convey the sounds, smells, and tastes of the intimate.[16] Drawing on fieldwork I want to consider how micro and very intimate practices intersect with the macro forces of supply chains and foodways—in many ways similar to what Ian Cook calls the "geographies of food: following."[17]

The argument I construct of necessity takes me away from feminist debates about representation. With many notable exceptions, much feminist critique, especially in cultural studies, has been mainly cultural in nature. While the politics of representation, especially in terms of the media, has of course advanced an important intervention, this has recently been challenged by the call to 're-materialize" social science research. In geography, as Sarah Whatmore and many others have argued, the very matters of life and the planet need to be rethought.[18] Whatmore argues that "cultural geographers have found their way (back) to the material in very different ways that variously resonate with . . . the most enduring of geographical concerns—the vital connections between the *geo* (earth) and the *bio* (life)."[19] In turn, Ben Anderson and John Wiley are quick to point out that they "reject outright the rhetoric of rematerialising or 'returning' as often erroneously and reductively equating materiality with 'ground,' with 'reality,' and with the 'social.'"[20] While I take heed of the many feminist calls from science and technology studies to go beyond culture, it is important to note that over decades different strands of feminist research have moved in and out of the material. As this collection attests, the history of thinking and figuring the materialities of the intimate has been a feature of feminist research in many different areas for a very long time.

There are, however, tensions emerging about how to conduct feminist studies when the object of analysis does not overtly concern either women or gender as central categories of concern. To put it bluntly, can there be feminism without women? This is, of course, the title of Tania Modleski's 1991 book, which summed up a certain stream of critique in the 1980s and 1990s.[21] In those fraught days the battle lines primarily separated "women" from "gender," with the accompanying division of certain articulations of "theory" (variously assorted under large and sometimes erroneous categories of "poststructuralism," "postmodernism," and "postfeminism") against "women's studies." For those younger than me, it may seem hard to believe that so many bloody words were waged in fights that often pitted women against women. For instance, in

Modleski's critique of Janice Radway's *Reading the Romance* (1984; a study of women readers that used ethnographic reader-response methods that at the time were groundbreaking in literary studies), it doesn't matter that the object of analysis was clearly women; the method itself was said to be masculinist. As Kate Kramer argued in her otherwise favorable review, Modleski's point was that ethnography is inherently flawed as a feminist tool because of "its disdain for any considerations of the unconscious of the subjects it seeks to describe."[22]

The question of how to conduct research on the materialities of the human and nonhuman may seem of a different order than the critiques about the true object of feminist concern, but issues remain in terms of a proper reckoning of a feminist project focused elsewhere than on gender or women. Modleski's attack on ethnography merits a pause, and a correction. I've argued elsewhere that researching the affective registers takes a toll on one's body and sense of self.[23] One can swallow others' emotions, breathing in and taking on together emotional registers. Taking from Georges Devereux, we begin to understand how profound this may be. In the 1960s Devereux founded an area of research called ethno-psychiatry. One of his most influential books is entitled *From Anxiety to Method in the Behavioral Sciences* (1967). Having first studied physics under the tutelage of Marie Curie, he trained as a psychoanalyst after completing a PhD in anthropology. His work remains important in the clinical treatment of mental disorders in non-Western countries and has also been influential in Gilbert Herdt and Robert Stoller's anthropological work on sexuality.[24]

Devereux's ideas on how methods within social science both protect and promote anxiety in the researcher are central to my ideas here. He argues that the singularity of the social sciences is that they deal with affective material: "the analysis of man's conception of himself." Of the three great revolutions— Copernicus, Darwin, and Freud—he remarks that, "It was easier to be objective about the heavenly bodies than about man as an organism, and the latter easier than objectivity about man's personality and behavior."[25] The trouble and, conversely, the great potential of the sciences of man (in particular, anthropology, sociology, and some branches of psychology) lies in "the difficulty in clearly distinguishing materials that come from outside (the subject, the field) and from inside (his or her own emotional reactions). The researcher has to struggle with these emotional reactions and anxieties. . . . The researcher is, in one way or another, the subject and object of the knowledge that he/she elaborates."[26]

This framing of ethnographic research more than responds to Modleski's critique that ethnography is a disembodied exercise in accounting for external realities. Indeed the inter- and intrasubjective encounters within ethnographic research are continually fraught with questions prompted with morphing boundaries. As I've written elsewhere, the problem is not that emotions are absent in the research process; it is that we need ways of being more precise about which emotions and which effects are at play, and for whom.[27] The stakes on these questions are ramped up when the focus is on the materiality of encounters that are in themselves often seemingly mundane.

The research that fuels this chapter builds on ways of thinking intimacy and globalization in a different voice, and is part of a larger project on how the relationships between taste (consumption) and place (production) are being rearranged in the globalized food system.[28] Here I want to explore taste in its various guises, in a conscious move away from the dominant sociological framing of taste as first and foremost, if not always, about social distinction. As with Whatmore's approach, I will employ a "self-conscious act of storying," what she calls "an envoy of the recuperation of 'materiality'"[29] or what I elsewhere call a rhizoethology of bodies.[30] This casts research as a wending through ethnographic observation and the stories told by different individuals. My concern here is what I am calling the interests of taste and place. Taking "interests" as both economic and social, and as also driven by curiosity, the etymology of "interest" refers us to the "inter-esse," and the materiality of the different forms of being in between. Isabelle Stengers, the Belgian feminist philosopher of science, uses the etymology of interest to talk "of bearing witness to the many aspects of entangled slow stories."[31] There are many interesting slow stories about the place of taste and the tastes of places. Telling them, listening to them, and passing them on are part of the process whereby we relearn to be interested in the material connections of taste and place. This is for me one aspect of a crucial ethics of eating, of taste and place.

Taste is, of course, so very intimate. Here I want to focus on how taste and tasting as material practices can also come to form communities: local places that are entangled in the global. I move from one particular organism, oysters, and follow the network they form. I chose oysters because at a micro level they are such a unique and intimate taste. Now seen as a luxury, and therefore associated with social distinctions of elitism, they have had a long history of being poor people's food. At a macro level, studying the farming of the Pacific *Crassostrea gigas* reveals the connections between community, economy, and a sense of locality within the global. The local that emerges out of these connections is not to be valorized for its own sake. It is as mundane as an oyster lying in its bed before it meets its fate to be savored and praised as a branded "Loch Fyne" oyster in swanky restaurant. Simply put, oysters are good to think with, and they make me think differently about mundane and grounded natures of taste attachments.

LOOKING AN OYSTER IN THE MOUTH

Oysters have been eaten since the beginning of mankind's time, or at least by those living by the sea. In Australia, oyster middens attest to Aboriginal appetites and good taste over tens of thousands of years. The poor and kings alike have partaken of their pleasures. In a visceral way, oysters divide people not so much into social classes—although the costs may divide ocean-dwellers from the landlocked—but into more primordial clusters. For instance, in her book *Consider the Oyster* (collected in *The Art of Eating*), M.F.K. Fisher describes how human oyster-eaters fall into three groups: "Those loose-minded sports who will eat anything, hot, cold, thin, thick, dead or alive, as long as it is *oyster*; those who will eat them raw and only raw; and those who with equal severity will eat them cooked and no way other."[32]

In general, I'd say that a taste for oysters polarizes humanity into two camps: for and against. I am firmly in the former and like them all ways, especially raw. There is the taste, of course—that briny, nectar, and zinky combination like no other. Then there is the texture — plump or fine and slightly gritty, or milky and unctuously smooth. But mainly it's the sensation, a thrill that passes through membranes reaching to the tips of my toes. There is something so unworldly and exciting about the feeling of live flesh on flesh, as oyster and human combine.

The world of food writing has become democratized through online forums, and I like the ways in which this opens up descriptions and ideas about how things taste. For instance, listen to this food blogger: "Eating raw oysters is a uniquely invigorating experience; a bit like battery-licking for grown-ups. It seems that we can taste the elements they contain: zinc, calcium, copper, iodine, magnesium. And no other food conjures up a physical feature of the Earth as strongly as a bracing, salty, tangy oyster: the essence of the sea in edible form."[33]

Like this writer I don't discriminate in my love of oysters. I have eaten Pacifics in the Pacific Northwest of North America, Sydney Rocks in the Rocks in Sydney, Belon in Paris, oysters from Bouzigues grown in the brackish water of the Thau, fine Irish ones in a Dublin pub, and of course the splendid Coffin Bay oysters harvested in the Eyre Peninsula in South Australia, home to some of the world's most isolated and pristine waters.

Years ago, conducting research on Scottish regionality, my gal and I stumbled upon Rogano's Oyster Bar in Glasgow, a celebrated institution. We sat in a thin sun among the cheerful Glaswegian crowd and feasted on oysters, swearing they were the best ever. "Where are they from?" we asked, ordering more. The Scottish lass answered without a hesitation. "Loch Fyne, they're the best."

So began our quest to find Loch Fyne, to which I'll return. But first let's consider the sheer strangeness of the oyster. James I, in the seventeenth century, is reported to have said that "he was a bold man who first swallowed an oyster."[34] The indeterminacy of these bivalves extends to their sex, which changes seemingly at whim. They start off as larvae and then evolve as spats. In her chapter "Love and Death Among the Molluscs," M.F.K. Fisher describes how:

> Almost any normal oyster never knows from one year to the next whether he is he or she, and may start at any moment, after the first year, to lay eggs where before he spent his sexual energies in being exceptionally masculine. If he is a she, her energies are equally feminine, so that in a single summer, if all goes well, and the temperature of the water is somewhere around or above seventy degrees, she may spawn several hundred million eggs, fifteen to one hundred million at a time, with commendable pride.[35]

The "true" or eating oysters are from the Ostreidae family, whereas the pearling oysters belong to the *Pinctada* genus. The two are not very closely related, so the chances of finding a pearl in your dozen au naturel are slim. Oysters breathe like fish, with gills. They also have a mantle lined with blood vessels (colorless blood); through which they take in oxygen and expel carbon dioxide. It's a little disconcerting to learn that they have a heart, and indeed it is a three-chambered affair. They have two kidneys, which act to purify the blood of waste products. They are part of what are called keystone species, which means that they benefit their ecosystems in a disproportionate ratio to their size and abundance. For a relatively small entity they do a lot of good. One oyster can filter up to five liters of water an hour. Rowan Jacobsen calls them "ecosystem engineers" that enable a "flourishing web of life."[36]

TASTE POSITIONS

Much of cultural theory laments the loss of the authenticity of taste and place. For instance, Albert Borgmann, an American philosopher, deplores the practices of consumption, which "displace things that have a life and a dignity of their own."[37] I take these terms literally: to think about how our tastes are displaced in our modern food system. My project takes a more productive turn, and I want to reframe how we understand consumption and production. Given half a chance, eating reminds us of the different ways we are connected: economically, globally, communally, and emotionally. As Jacobsen says of oysters, "the comparison for oysters is not a taste or another food but always a place . . . the seacoast."[38] Oysters give us a taste of "somewhereness."[39] Taste and tastes are forms of attachments: to the strangers who grow or produce food, to friendships, which are made and remade in commensality and sharing, to the soil and water, the factories, ships, oyster leases and abattoirs. Taste ties us to history just as it does into various economic relationships.

For instance, you can't understand the flourishing seafood industry in the Eyre Penisula, an isolated and rugged coast in South Australia that faces Antarctica, without understanding the connections forged by history, migration, and the clusters of settlement that form and reform around water, harbors, and land. The first occupants of the Eyre were the Nanuo Aboriginal mobs who would move between the sea and the land depending on the time of year. The early white settlers were "peasant fishermen" from Ireland, fishermen and farmers, who according to a retired fisherman I talked to, had an ethos of farming the sea rather than strip-mining it. This man deplored current practices of overcrowding the oyster leases, which leaves nitrates in the sea floor.[40] He told me how the present-day oyster industry stands on a history of the disastrous effects of dredging, which killed off the native oysters. He recounted other disappearances, such as that of the Chinese who worked the land and the water, and left traces of their culture on the Anglo-Celt population. Standing in the bar he pointed at a younger fisherman who wore the "lucky 8"—the coiled snake conferring luck within Chinese culture.

I want to frame taste as "those things that hold us together," to quote the French sociologist Antoine Hennion. "Taste," he says, "is what links us, constrains us, holds us and what we love, what binds us, that of which we are a part."[41] "Taste is a problematic modality of attachment to the world."[42] This description begins to capture the way taste acts as a connector between history, place, things, and people. It contrasts with dominant sociological understandings of taste, which have tended to be rather utilitarian in their considerations of what taste does in society. Taste is often considered a fundamental marker of one's position in society, which in Pierre Bourdieu's terms serves to reproduce inequalities.[43] Taste, for Bourdieu, is the mechanism through which hierarchies are affirmed: "taste classifies the classifier." While, of course, Bourdieu's theoretical points about tastes as the exercise of judgment was based on his copious questionnaire data, which queried people about their eating activities as well as other leisure practices, in his terms taste became reified as social casting and moved away from the sensory realm of tasting. The frame of class remains crucial, but it is important to remember that the ability to taste is not the province of the few (albeit the leisure and the money to let your tastes roam widely is increasingly seen as the domain of the rich). In their insightful article on taste and feminism, Alison and Jessica Hayes-Conroy examine the intersection of the labeled body, which is, as they say, affected by the labeling but "such affecting occurs in the visceral realm—the realm in which the whole molecular ensemble of the minded body feels the world, the realm from which life processes and events precipitate and hence in which political activation *materially unfolds* (or fails to)."[44]

Their argument joins those of other feminists such as Bev Skeggs[45] and Lisa Adkins and others[46] who have worked hard to render the place of gender more central to a Bourdieusian framework. But to only focus on those bilateral operations of taste-connections and connotation misses some of the wider mysteries that taste discloses. Perhaps ironically, taste in Bourdieu's sense, and in those of many others who take from him, is treated as a property, which some own and use over others. Taste here veers off into the abstract. Which tastes? What sensations? If we think about taste as different types of attachment, we can begin to move beyond focusing just on humans and instead consider the different connections of human, nonhuman, technology, history, and culture. As Elizabeth Silva argues, instead of obsessing on the homologies set up in consumption, we might turn to the forms of appropriation and their effects, in which individuals (and most particularly women) operate as "conscious agents of change in the sphere of intimacy."[47] To think about how the world is rearranged through the material appropriation of new tastes, and the discovery of tasting anew, opens the doors to radically different visions of what intimacy means to whom or what. What I particularly like about this approach is that it foregrounds the *mattering* of taste. As Hennion writes: "Taste is not an attribute, it is not a property (of a thing or of a person), it is an activity. You have to do something in order to listen to music, drink a wine, appreciate an object. Tastes are not given or determined, and their objects are not either."[48]

Taste is something you have to do. Hennion says that tasting is a "perplexed" activity, which is to say it makes you "on the lookout for what it is doing to you, or what it is doing to others. You have to lend attention when you taste, if you are to fathom the taste. One 'observes oneself tasting.'"[49]

The body that tastes is also a body in training. It's clear that through learning to taste people also acquire a different sense of their bodies and of the world, as well as of the different ways of describing how they fit within a complex web of things and tastes. In other words, through taste our bodies learn to be affected, and to reflect upon how and what affects us. Bruno Latour writes that learning to be affected and "training the nose" means "acquiring a body [in] a progressive enterprise that produces at once a sensory medium *and* a sensitive world."[50] In a related study of professional tasters, Cécile Méadel and Vololona Rabeharisoa demonstrate that "the body is the main mediator between product and taster. . . . The production of a sensitive body is the basic operation in the process [of tasting]."[51] As the body becomes a sensitized medium, relations between different parts of the body vibrate differently in tandem with different aspects of the thing being tasted. This activity remakes the world, makes up new worlds. Latour talks about how odor training changes the relationship between words and their objects. To talk about tasting oysters as battery-licking, as that blogger does, is to think differently about how body parts act and react to other non-human bodies.

It's a lovely idea that taste carries worlds. But in practical terms, how does one go about researching such a seemingly ephemeral notion? As you can see, taste covers so much—from the cultural and the economic to the personal and the collective. My approach is to bring together an interdisciplinary range of ideas and theories and then to turn downward toward local case studies—to use the theories to catch meanings and then to land them in local studies. At this level I proceed with a mixture of methods and construct case studies from different documents as well as interviews and ethnographic observation. But as I flagged at the outset of this chapter, the encounters are deeply embodied—and in fact the ethnographer's body becomes part of the processes of attunement, whereby bodies are brought into dialogic relationships. This type of research requires methodological sensitivity to the different forms of tacit knowledge and enskilment—for instance, the ways in which fishermen know fish,[52] the embedded knowledge of fishing communities,[53] and the expertise of ecologists and fisheries managements.[54] This is what Harbers, Mol, and Stollmeyer call "*involved* description: ethnographic work that looks for contrasts, sets up differences, and seeks for what one practice might learn from another."[55]

Let us now return to the story of Loch Fyne, a small but instructive knotting together of people's lives, livelihoods, oysters, and tastes past and present. The Loch Fyne Oyster Company lies at the top of the loch in the little village of Cairndow in the west of Scotland, just beyond the stop called "Rest and be thankful." The hand of the Highland Clearances is everywhere—from the limited infrastructure and sparse population to the beautiful wilderness that is a testament to the lack of people, modernity, industry, and money. Following the failed Jacobite rising in 1745, the clans began to disintegrate, as did the system of obligations and responsibilities that had long held Gaelic Scotland together. The vast estates began to be run by absentee landlords, or rather by their factors while their masters "lorded" it around London. Sheep were introduced because of the high price of wool. The English-bred landlords also brought in workers from the south and rendered the small tenants redundant. For a time the locals eked out a living by harvesting kelp, but as the rule of the landlords and their factors became ever more draconian, they were forced to emigrate in poor circumstances to faraway places: to the "pink bits" that were the British Commonwealth. It's estimated that between 150,000 and 250,000 were evicted because they got in the way of the sheep.

It is not an overstatement to say that Highlanders still live the legacy of that time. The Highlands are a stark example of the often vexed interrelations of people, animals, and environment: it's a habitus, if you like, but one which includes the elements, weather, climate, a history of empty stomachs, and fear and respect for the sea and for nature's reproductions. There is always a thin but visceral layer of remembrance: that some humans, in their greed, placed sheep or cattle higher, on a strange hierarchy, than other humans. And of course some of those same people came to Australia and reenacted another hierarchy over the Aboriginal inhabitants, placing them on par with the flora and fauna of this bewildering new land.

Lack of employment and the sea is key to the story of the Loch Fyne Oyster Company, as is an enduring sense of responsibility to community. This is where Johnny Noble enters the picture. Noble was the son of an industrialist, Sir Andrew, who had made the family fortune in armaments and warship works in Newcastle. When his father died Johnny inherited the estate and was none too pleased to find that his father had heavily mortgaged the place. From all accounts Johnny was a good man, and he couldn't stand to see the people who worked the estate for generations chucked out.[56] Pondering what to do, he walked his seemingly bankrupt lands down to the loch. There he fell into conversation with Andrew Lane, a marine biologist and fish farmer. They hatched a plan to farm oysters. They bought the Pacific oyster spats because Pacifics grow faster. In 1980 they put a table outside on the road next to the Loch and started selling oysters and kippers. Lord knows how, but the enterprise took off. When they started farming oysters, the average consumption of oysters in the entire

United Kingdom was 5,000 a year. Now, as the largest supplier in the country, they sell the majority of the 30,000 dozen eaten each year. Noble and Lane were smart, and thanks to Johnny's flair and connections and Andrew's careful tending of the stock, they were soon supplying all the top restaurants in the country, as well as far-flung parts of the former Empire, such as Hong Kong and Singapore.[57]

Tragedy struck Loch Fyne when Johnny died in 2002 at the early age of sixty-five. He had done what he set out to do, and had provided for the local people he cared about. He also ensured that money went back into the community for the "relief of poverty, the advancement of education and the protection of the environment particularly within Scotland and the area around the head of Loch Fyne."[58] With the death of Johnny the future looked grim. Then another player entered the picture. Baxi Partnerships is a trust set up to help employees buy their company. David Erdal, the chairman at the time, helped the employees set up a deal whereby they could buy Loch Fyne Oysters with the help of the Royal Bank of Scotland. The deal worked, and the employee/owners have since expanded the business further. It's a lovely story of one community growing and trading in on a globalized taste for oysters that has enabled the community to grow, connecting past with present and future. For instance, the manager of the oyster bar is Christine MacCallum, who is the wife of the shepherd on the Ardkinglas estate, who is himself the son and grandson of shepherds before him. In her words you hear the defiance and pride: "We had been subjected to ridicule for many years as the idea of growing oysters in Loch Fyne was mad and the idea of having an oyster bar anywhere outside of London was, too."[59]

ECONOMIES OF TASTE ATTACHMENTS

On one level, taste opens new worlds, and by lending ourselves to taste in its profundity we experience the very thingness of what we are eating, drinking, tasting. But how can we reconcile the present-day system of supply and demand, of consumers and producers, tightly regulated by price and profit, with these ideas? Studies on the production side of eating and drinking tend to frame small producers, consumers, and suppliers as passive in the face of big retail. Dominant theories of economics tend to be unilateral in their vision (although as I discuss later, this is being disrupted by the important intervention of J.-K. Gibson-Graham). Consumers are to be manipulated; conversely, they are attributed a power quite disproportionate to most circumstances. The convoluted relationship between consumers and the economy is much more complex: desires, structures, and financial clout all complicate any causal explanation of how people and economies are connected. Within this web there are further twists in the interrelationships of humans and nonhumans, technology and nature.

The ideas of Michel Callon, a sociologist of economics, offer an interesting take on how to think about the economy as sets of entanglements. Callon, along with Latour, works at the Ecole des Mines in Paris where actor-network theory was originally developed. Originating in the anthropology of science and technology, they study networks or assemblages of things—human, technical, nonhuman—and how they interrelate. For instance, in a study of the scallop industry he describes "a network of relationships in which social and natural entities mutually control who they are and what they want."[60]

What I like about this approach is that it directs attention to "small events, little gaps, local innovations, limited reconfigurations."[61] Against the passivity that lurks in descriptions of both consumers and suppliers, this perspective offers agency and action. It's a special take on action and agency akin to things exerting pressure on each other in particular ways rather than some all-conquering notion of control and power. Nature, technology, and humans make markets as entanglements in which we humans are but one factor.

So with these ideas in mind, I want to consider how taste economies might shed new light on the intimate entanglements of people, places, and markets. I'll start by considering once more the oyster. Oysters obviously like Loch Fyne. While not the native variety, the hardy Pacific gigas took to the low-saline waters, where the ocean mix has lost some of its brininess. The Scottish lochs are ancient fingers pushing into the land, which makes the seawater change as it enters into the circulation system of the loch. While the weather is often less than kind to humans, to an oyster it's a fine place to be an oyster. Scientists have recently discovered that oysters are gregarious. They like to put down their feet where other oysters hang out. However, adult oysters are cannibals who, with the starfish and weird entities that burrow into their shells, pose the greatest threat to young oysters. So wee larvae, cognizant (on some level) of the danger posed by their grown-ups, nonetheless prefer to be together. You can hear the excitement of the scientists as they describe how: "Oyster larvae make a life or death decision when they get their one chance to select where to attach themselves to the bottom. Our research shows that oyster larvae are willing to risk predation by adult oysters to cash in on the benefits accrued by spending the remainder of their lives among a large number of their species."[62]

Oyster lovers, scientists or not, are such that our interest in the bivalve can veer toward anthropomorphism. However, I like to think of this interest as attachment. This sense of attachment goes far beyond the feeling one might have toward an inanimate commodity. And this is not just because they are animate, and because they animate our appetites and bodies. It is because—and this is especially true of the Loch Fyne story—the ties between consumers, suppliers, environment, history, and the bivalve are so close. This is an artisanal sea ecology that, as Tim Ingold argues, arises from "an active and mutually constitutive engagement between organisms and their environments, and in no way precede[s] that engagement."[63] Or in Gisli Palsson's wonderful ethnography about enskilment at sea, humans and the ocean become attuned in symbiotic ways.[64]

One way to think of this is to turn the economic term of externalities on its head. In traditional economic theory, externalities are all those costs that producers do not have to—and therefore do not—include in their costings. Callon explains: "Economists invented the notion of externality to denote all the connections, relations and effects which agents do not take into account in their calculations when entering into a market transaction. If, for example, a chemicals plant pollutes the river into which it pumps its toxic products, it produces a negative externality."[65]

Callon goes on to describe positive externalities as well, but he goes well beyond the binary implied by good and bad externalities to his interest in the ways in which there are always overflows—of information, interest, money, etc.—that then foreground or perform what he calls hybrid forums where actors and their interests are in constant fluctuation. This is a variation on what Latour calls "attunement."[66] As Anderson and Wylie explain, "An attunement [is] how heterogeneous materialities actuate or emerge from within the assembling of multiple, differential, relations and how the properties and/or capacities of materialities thereafter become effects of that assembling."[67]

In the case of Loch Fyne Oysters, there are numerous externalities at play, which make us think differently about the assemblages formed in the "in between" of taste and place. As I've said, the oyster itself seems to like the loch. Part of the money provided by the sale of oysters is versed into the Loch Fyne Trust to care for the loch. This then aids the oyster to do its own work of cleansing—as we saw earlier, it is a keystone species, and if allowed to, it does a lot of good. A symbiotic relationship begins. Probably unknown to the oyster, humans too have had a long interconnected relationship with the environment oysters call their home. Gathering kelp, harvesting the peat saturated with seawater, gathering mollusks, and fishing have been a way of life interrupted only when the greed of some men turned attention away from the sea and onto the land to raise sheep. Through the Loch Fyne Trust, groups like Here We Are have been set up as a resource that they call an interface between the locals and the incomers, the young and the old. As Here We Are says on its Web site, the objective is "to explain how a rural community, while loyal to its roots, makes its way in the modern world. Its subject matter is people in a place."[68]

History is certainly an externality that may be only tangentially associated with the oyster as a commodity. But the fact remains that it changes the relationship of people to the oyster, morphing it into a form of attachment rather than a pure economic exchange. If we then turn to the story of the laird looking to care for his community, we enter into the scene that economists might more readily recognize as "economic." But while Johnny Noble made money from his enterprises, the obituaries and stories about him portray a man not particularly interested in furthering an individual economic gain. He was, of course, motivated by interest, but self-interest or altruism don't quite capture the ways in which his business spread over land, sea, oysters, fish, people, and history.

While we could call this outlook feudal, it is more correctly a profound sense of responsibility, which can be seen as a positive externality—a radically deep attunement of bodies and histories. The multiple forms of attachment within the community, which come to operate as a kind of ecosystem, impart a form of assurance. That assurance was traded upon, in the case of the Loch Fyne oyster enterprise, when Johnny Noble died and the possibility was offered that the employees could own their own company. In sociological terms, you could call it the conversion of cultural capital into economic capital. While many might contemplate how good it would be to have a stake in their own company, few succeed. As David Erdal, the man who brokered the deal, explains, "Employee ownership gives people more satisfying working lives than they ever thought possible."[69] The power to think beyond "what they ever thought possible" can be included in the long list of positive externalities associated with Loch Fyne Oysters.

Instilling a taste for oysters in this case builds on the interests, or in the etymological sense, the inter-esse, of beings and being in the "in between" of place and taste. I want to briefly turn to another example of how oysters may grow communities. Cowell is a small town on the other side of the Spencer Gulf, and down from Whyalla, on the Eyre Peninsula on the wild side of the state of South Australia. This is mining country, which grew out of copper explorations in the 1850s. Mining communities are now sutured into global multinationals—

Whyalla is a one-industry town dependent on One Steel, previously a division of BHP. Iron ore is mined around the Eyre, and you can see the tops of mountains that have been lopped off. The red dust covers everything. It mars the beauty of the outback and the sea, but it was iron ore and the sales to the Chinese that protected Australia from the full extent of the global economic meltdown.

When the ferry is running, Cowell is the first stop on the Eyre Peninsula across from Walleroo. However, the traffic for years has been generally the other way around as people leave toward the east. As with so many small communities built on farming, the town was losing its young people to larger towns or to the mines inland. In 1991 the local oyster growers approached the Cowell Area School with a plan to foster skilled workers in the oyster farming business—to urgently try to interest young people in oysters as a way of stopping the out-migration and to further the future of the local oyster business. They donated two hectares of oyster leases in Franklin Harbour, a secluded bay rich in sea grass and algae. For a while the aquaculture course worked beautifully and served as a beacon to rural schools in places as far afield as Hawaii and Idaho. However, by the beginning of 2009 it was in pieces, with DECS (Department of Education and Child Services, South Australia) threatening to close it down.

It seems that oysters and the taste attachments they could forge had become the property of a few. According to the new principal[70] it had become a "hobby" and no longer was seen as a public and community investment. Some of the negative externalities they faced included increased competition from other schools turning to aquaculture; paradoxically, the huge rise in oyster farming on the Peninsula may have served to close down rather than open up opportunities. Fifteen years ago there were very few oyster farms, and now there are more than 400. The industry has become highly professionalized and water is tested several times a day to ensure quality. This makes sense when oyster producers such as Pristine Oysters, located in Coffin Bay, ship two tons of oysters a week to Dubai, Hong Kong, and elsewhere.

When I went to visit the school in 2010 there were signs of new life. A banner outside the school proudly proclaimed their revitalized aquaculture program and thanked their sponsors, including Turner Aquaculture and BST (oyster farming equipment specialists, who coincidentally sell to Loch Fyne Oysters). Jan Potter, the incredibly hardworking principal, led me around the back of the school and introduced me to Mark, an enthusiastic young man who had returned to the school, where he did his aquaculture training, after having toured the world—including a stint working at Loch Fyne. They showed me a fascinating array of tanks, which teachers, parents, pupils, and other volunteers had painstakingly scrubbed and fixed. Someone had come up with the idea to put windows lower down in the large oyster tanks so that the "littlies" could see what was happening with the oysters. Every school year (or grade) has its own projects—from growing the local marine habitat and farming sharks and barramundi to growing the spats. Alongside the adults, they have a hand in the maintenance of the equipment and the leases. And of course there is the hard work of tending to the oysters, as well as grading and cleaning them.

In terms of what we might call the positive externalities of this business, what seems to be emerging is an assemblage of local growers, such as Turner Aqua, as well as a network that connects with Tasmania, where two growers donate spats. The network also connects the schools with buyers in Japan, who love buying oysters harvested by schoolchildren. Crucially, the oyster scheme is now treated as a business and not a hobby. It has become, in Latour's terms, an attunement of community, oysters, and environment. Jan Potter has put in place a solid business plan that should see that the returns pay for Mark's wages. But equally important are the ways in which oysters now play a part in the academic curricula in the school, from business studies, food technology, and art and culture to, of course, marine science. All of these factors are instilling a taste for oysters and a desire to learn about their taste and place in their area, and the school has become a microcosm of the wider ecology in Franklin Harbour. As we walked among the permaculture gardens fed by the recycled hydroponic nursery, I still was curious. "But do the young kids like the taste of oysters?" Mark laughed and said that when they were shucking, the kids popped oysters in their mouths as if they were sweets.

GLOBAL INTIMACY AND FEMINIST RESEARCH

I will now turn to draw out the import and consequence of such research—which for want of a better term we might call taste economies. It aims to be quite radically interdisciplinary in that, working in the inter-esse—in the "in between" of being and beings, of taste and place—it disrupts boundaries and reforms the object of study. This has important consequences for (some) feminist research. Of course, many of the human actors are women, and it matters that these women are seen and that the gendered dimensions of taste production are recognized for what they are—a huge factor in the production and preparation of food around the world. But what matters for the feminist approach I am trying to propose and embody is in the quality of the accounting and the ways in which we account for different attachments and connections. We might call this a feminist attunement to the little details so crucial to adequately understanding the big picture. This attention is nothing new and characterizes the work of feminists working across areas that are too many to mention but notably include literature,[71] cultural studies,[72] and media studies;[73] younger scholars such as Jane Simon[74] have painstakingly returned to the question of detail in a new vein. What is also new, perhaps, is a critical impatience with the dictates of fashion and theory. Not for nothing were feminists back in the 1990s wary of the shifts to ever higher plains of theory that often clearly announced the end of one thing after another (including women, it sometimes seemed).[75]

Along with others, I hope to advance the move away from the dominant negative tone of constructionism. For some of us it has become increasingly important as a matter of life and thinking to try to seek out new aspects of intricate arrangements of animate and nonanimate life. As J.K. Gibson-Graham argues strongly, what is important now is first to practice "an experimental attitude toward the objects of our research"; second is "an orientation toward collective research practice involving academic and non-academic subjects."[76] I'll expand on the second point, but I want to dwell for a moment on the exigency of an experimental stance or attitude. Graham-Gibson takes her lead in this from Eve Sedgwick's question: "What if we were to accept that the goal of theory is not to extend knowledge by confirming what we already know . . . what if we asked theory instead to help us see openings, to provide a space of freedom and possibility?"[77] It was Sedgwick's remarkable critique of what theory thinks it knows, in her introduction with Adam Frank to an edited volume of the writings of Silvan Tomkins,[78] that set me off on my own determined, oblique path to study the wide workings of shame some years ago. It is perhaps surprising that an American clinical psychologist writing in the 1940s and 1950s should galvanize economic geographers and a literary critic. But maybe not. As Sedgwick and Frank write of Tomkins's effect: "In a sodden landscape of moralistic or maudlin *idées reçues* . . . Tomkins' formulations startle."[79]

While this is not the place to rehash my own infatuation with Tomkins (and with Sedgwick), suffice it to say that reading him/them was like battery-licking (to again quote the oyster blogger). For instance, Tomkins's description of how shame is intimately entangled in interest hauled me in: "that the pulsations of cathexis around shame, of all things, are what enable or disenable so basic a function as the ability to be interested in the world."[80] Interest in Tomkins's sense can be seen as the range of stimuli that constantly hit us, move around and through us, and most important of all connect us . . . to people, to feelings, to memories, to water, to land, to oysters, to women, to parts of the world near and far. In this way I learned to read across a very wide assortment of ideas and theories (albeit few of them in my "home" disciplines), not in the judgmental way (in which I was trained—"what are the conditions of possibility, of truth and falseness") but in terms of "wow, isn't that interesting?"[81]

It's important that this stance (or what we might call an attitude in Foucauldian, or style in Deleuzian, terms)[82] is not newness for the sake of newness itself. It is not particularly feminist to declare that "a certain form of critical spirit has sent us down the wrong path," as Bruno Latour recently declared in his review of why critique has run out of steam.[83] And, in fact, the mini-industry now following Latour might do well to turn their critical attention to what it is that capitalized ventures such as ANT actually *do*. For as Deleuze so patiently reminded us, concepts and ideas must be generative or else they no longer deserve their appellation as such.

Traveling alongside the many actors, human and nonhuman, my tale of oysters, told from the stories of so many, may allow for an oblique but hopefully useful conception of the field of taste as it is being formed and re-formed in localities and regions seemingly outside of the mainstream. As Callon recounts, "The growing complexity of industrialized society, [and] a level of sophistication due in large part to the movement of technosciences, are causing connections and interdependencies to proliferate."[84] But to extend his thoughts, we need to foreground gender within the materiality of hybrid forums. The complexities cannot be solved or resolved by (the often recognized male) experts working on their own alone in labs. As Gibson-Graham so clearly urges, experts and nonexperts must continually interact. Research as a clustering of objects, different experts, and contexts—a feminist ecology of research—brings together very different actors to work at understanding large-scale issues through grounded and localized action. Those not from the university share their interests, knowledge, and attachment to particular facets, while in turn some of our academic interests rub off on them.[85] In terms of the research I have been doing in remote areas, this plays out in long and fascinating conversations with fishermen, with restaurant owners trying to promote local gourmet food, with the owners and managers of pubs, with oyster growers and sellers, with principals trying to invigorate their schools and communities, and with representatives of different industry and government bodies.

And of course, there is the eating of countless yet all distinct oysters, each creating different sensations of flesh on flesh. Together this assemblage can cast forward ideas about how to further taste economies deeply entrenched in communities yet reaching far beyond to tantalize tongues around the world.

NOTES

My thanks to Clif Evers, Jane Park, Kane Race, and Shvetal Yvas for their comments on earlier drafts, and to Geraldine Pratt and Victoria Rosner for their close and generous editorial advice. Thanks too to Sarah Donald, who shares a love of oysters.

1. Harold Innis, *The Cod Fisheries*.
2. Innis, *Cod Fisheries*, 52
3. Pierre Bourdieu, *Distinction*.
4. The issue of obesity is mired in often unspoken assumptions about class and gender. A team of researchers from a transdisciplinary base recently set off a heated debate in the pages of the *International Journal of Epidemiology* (2006). Led by Paul Campos, a legal scholar, they argue that the global obesity epidemic is a moral panic, which, as they say, "often displaces broader anxieties about changing gender roles." They cite as evidence an advertisement that ran in the *Washington Times*: "30 years of feminist careerism . . . with most mothers working, too few adults and children eat balanced, nutritious, portion-controlled home-cooked meals." The ad goes on to predict that 50 percent of Americans will develop diabetes, "creating a medical and financial nightmare, likely to crush our healthcare system" (Campos et al., "The Epidemiology of Overweight and Obesity").
5. Elspeth Probyn, "Feeding the World," 55.
6. Michael Winter, "Embeddedness, the New Food Economy and Defensive Localism."
7. Fabio Parasecoli, "Postrevolutionary Chowhounds," 38.
8. Julie Guthman, "Can't Stomach It."
9. Julie Guthman, "Fast Food/Organic Food," 54.
10. Julie Guthman and Melanie DuPuis, "Embodying Neoliberalism," 427–28.
11. Lauren Berlant, "Slow Death (Sovereignty, Obesity, Lateral Agency)," 756.
12. Ibid., 773.
13. Guthman, "Neoliberalism and the Making of Food Politics in California."
14. Berlant, "Slow Death," 780.
15. Adrienne Rich, *Bread, Blood, and Poetry*, 17.
16. Geraldine Pratt and Victoria Rosner, "Introduction: The Global and the Intimate."
17. Ian Cook et al., "Geographies of Food: Following," 656–66.
18. Sarah Whatmore, "Materialist Returns." If Whatmore is careful to distinguish herself from cultural geography, I might underscore that I am not a geographer, although my interest has always been close to that widely defined field in terms of seeking out ways of understanding and describing the material and the tangible aspects of life.
19. Whatmore, "Materialist Returns," 601.
20. Ben Anderson and John Wylie, "On Geography and Materiality."

21. Tania Modleski, *Feminism Without Women.*

22. Kate Kramer, "Tracing the Vanishing Woman," 161.

23. Elspeth Probyn, "Eating for a Living."

24. Alain Giami, "Counter-transference in Social Research."

25. Georges Devereux, *From Anxiety to Method in the Behavioral Sciences,* 3.

26. Giami, "Counter-transference."

27. Elspeth Probyn, "Glass Selves."

28. Elspeth Probyn, Taste and Place: The Transglobal Consumption and Production of Food, Australian Research Council Discovery Project, 2009–2011. I would like to thank the Institute of Geography at the University of Edinburgh for hosting me as a visiting professor in 2008, which enabled me to conduct research in Scotland on taste and place.

29. Whatmore, "Materialist Returns," 601.

30. Elspeth Probyn, "Eating for a Living."

31. Isabelle Stengers, "Another Look."

32. M.F.K. Fisher, *Consider the Oyster,* 125.

33. "Eat Oysters," www.eattheseasons.co.uk/Archive/oysters.htm (accessed 29 October 2008).

34. Rowan Jacobsen, "The Taste of an Oyster."

35. Fisher, *Consider the Oyster,* 125.

36. Rowan Jacobsen, *A Geography of Oysters.*

37. Albert Borgmann, "The Moral Complexion of Consumption," 421.

38. Jacobsen, *A Geography of Oysters,* 2.

39. Jacobsen, *A Geography of Oysters,* 3.

40. "Peter" is a long-term resident of the area. His father was a farmer, but he decided to go to sea, as had his forebears in Ireland. He still owns a crayfish boat and license, the latter worth some $5 million dollars. (Interview conducted 12–13 September 2009)

41. Antoine Hennion, "Those Things That Hold Us Together," 103.

42. Antoine Hennion, "Pour une pragmatique du gout," 1.

43. Bourdieu, *Distinction.*

44. Alison Hayes-Conroy and Jessica Hayes-Conroy, "Taking Back Taste," 462.

45. Bev Skeggs, "The Making of Class and Gender Through Visualising Moral Subject Formation."

46. Lisa Adkins and Bev Skeggs, eds., *Feminism After Bourdieu.*

47. Elizabeth Silva, "Gender, Class, Emotional Capital and Consumption in Family Life," 154.

48. Hennion, "Pour une pragmatique," 101.

49. Ibid.

50. Bruno Latour, "How to Talk about the Body?" 2–3.

51. Cécile Méadel and Vololona Rabeharisoa, "Taste as a Form of Adjustment Between Food and Consumers," 242.

52. Gisli Palsson, "Enskilment at Sea."

53. Kevin St. Martin, "The Impact of 'Community' on Fisheries Management in the U.S. Northeast"; Holly Hapke, "Petty Traders, Gender, and Development in a South Indian Fishery."

54. Thomas McGuire, "The Last Northern Cod," 1997.

55. Hans Harbers, Annemarie Mol, and Alice Stollmeyer, "Food Matters," 207.

56. Craig McQueen, "The Unlikely Friends Behind Scots Oyster Firm That Became International Hit."

57. R. Hoar, "Coming Up Fast."

58. Loch Fyne Oysters.

59. Craig McQueen, "Unlikely Friends."

60. Michel Callon, "Some Elements of a Sociology of Translation."

61. Michel Callon, "Why Virtualism Paves the Way to Political Impotence," 3.

62. Mario N. Tamburri, Richard K. Zimmer, and Cheryl Ann Zimmer, "Mechanisms Reconciling Gregarious Larval Settlement with Adult Cannibalism."

63. Tim Ingold, "Evolutionary Models in the Social Sciences," 242.

64. Palsson, "Enskilment at Sea."

65. Michel Callon, "Actor-Network Theory," 5.

66. Bruno Latour, *Reassembling the Social*.

67. Anderson and Wylie, "On Geography," 320.

68. "Here We Are."

69. Ibid.

70. Jan Potter, the principal of Cowell Area School, previously worked at another school in the Eyre and before that in the Aboriginal territory known as the APY Lands (governed by the Anangu Pitjantjatjara Yankunytjatjara Council). (Interview conducted 11 September 2009)

71. Naomi Schorr, *Reading in Detail*.

72. Meaghan Morris, "A Gadfly Bites Back."

73. Tania Modleski, "The Rhythms of Reception."

74. Jane Simon, "An Intimate Mode of Looking."

75. I am thinking here of the late 1980s and the heated debates and battles waged over postmodernism. Some landmark interventions by feminists served to bring some semblance of common sense and procedure to questions that threatened to swirl into hyperbole. See, for instance Meaghan Morris's introduction to her *The Pirate's Fiancée*, where she calmly lists feminist work that was engaged with postmodernism without the attendant male angst, or Linda Nicholson's edited volume, *Feminism/Postmodernism* (where, incidentally, I began my interest in locality).

76. J.-K. Gibson-Graham, "Diverse Economies," 628.

77. Eve K. Sedgwick and Adam Frank, "Shame in the Cybernetic Fold," 19.

78. Sedgwick and Frank, "Shame in the Cybernetic Fold."

79. Ibid., 4–5.

80. Ibid., 5.

81. Probyn, *Blush*.

82. For instance, many years ago, following Foucault's injunction, I tried to formulate "sexing the self" (Probyn, *Sexing the Self*). Later I attempted an argument indebted to Deleuzian style (Probyn, *Outside Belongings*).

83. Latour, "Why Has Critique Run Out of Steam?," 231.

84. Callon, "Why Virtualism Paves the Way to Political Impotence," 12.

85. I have, for instance, learned so much from being involved in the South Australia Food Science cluster led by Dr. Tim Rayner, where scientists, social scientists, and "end-users" engage in an incredibly constructive (and sadly often unusual) conversations.

BIBLIOGRAPHY

Adkins, Lisa, and Bev Skeggs, eds. *Feminism After Bourdieu*. Oxford and Boston: Blackwell, 2004.

Ambjörnsson, Fanny. "Talk." In D. Kulick and A. Meneley, eds., *Fat: The Anthropology of an Obsession*, 109–20. New York: Penguin, 2005.

Anderson, Ben, and John Wylie. "On Geography and Materiality." *Environment and Planning A* 41 (2009): 318–35.

Audra, L. "Learning to Taste: The Stages of Tasting." 1997. www.suite101.com/article.cfm/wines/108 (accessed 18 September 2008).

Berlant, Lauren. "Slow Death (Sovereignty, Obesity, Lateral Agency)." *Critical Inquiry* 33 (Summer 2007): 754–80.

Borgmann, Albert. "The Moral Complexion of Consumption." *Journal of Consumer Research* 26, no. 4 (2000): 418–22.

Bourdieu, Pierre. *Distinction: A Social Critique of the Judgement of Taste*. Cambridge: Harvard University Press, 1984.

Callon, Michel. "Actor-Network Theory—the Market Test." In J. Law and J. Hassard, eds., *Actor Network Theory and After,* 181–95. Oxford: Blackwell, 1999.

——. "Some Elements of a Sociology of Translation: Domestication of the Scallops and the Fishermen of St. Brieuc Bay." In John Law, ed., *Power, Action, and Belief: A New Sociology of Knowledge?*, 196–223. London and New York: Routledge, 1986.

——. "Why Virtualism Paves the Way to Political Impotence: A Reply to Daniel Miller's Critique of The Laws of the Markets." *Economic Sociology* 6, no. 2 (2005): 1–15.

Campos, Paul, Abigail Saguy, Paul Ernsberger, Eric Oliver, and Glenn Gaesser. "The Epidemiology of Overweight and Obesity: Public Health Crisis or Moral Panic?" *International Journal of Epidemiology* 35 (2006). http://ije.oxfordjournals.org/cgi/reprint/dyi254v2 (accessed 6 June 2008).

Cook, Ian, et al. "Geographies of Food: Following." *Progress in Human Geography* 30, no. 5 (2006): 655–66.

Devereux, Georges. *From Anxiety to Method in the Behavioral Sciences*. The Hague: Mouton, 1967.

"Eat Oysters." www.eattheseasons.co.uk/ Archive/ oysters.htm (accessed 29 October 2008).

Erdal, D. *Local Heroes: How Loch Fyne Oysters Embraced Employee Ownership and Business Success*. London: Penguin, 2008

Fisher, M.F.K. *Consider the Oyster* (1941). In *The Art of Eating*. New York: Wiley, 1954.

Giami, Alain. "Counter-transference in Social Research: Georges Devereux and Beyond." London School of Economics, Papers in Social Research Methods Qualitative Series no. 7, 2001.

Gibson-Graham, J. K. "Diverse Economies: Performative Practices for 'Other Worlds.'" *Progress in Human Geography* 32, no. 5 (2008): 613–32.

Guthman, Julie. "Can't Stomach It: How Michael Pollan et al. Made Me Want to Eat Cheetos." *Gastronomica* 7, no. 3 (2007): 75–79.

——. "Fast Food/Organic Food: Reflexive Tastes and the Making of 'Yuppie Chow.'" *Social & Cultural Geography* 4, no. 1 (2003): 46–58.

——. "Neoliberalism and the Making of Food Politics in California." *Geoforum* 39, no. 3 (2006): 1171–83.

——, and M. DuPuis. "Embodying Neoliberalism: Economy, Culture, and the Politics of Fat." *Environment and Planning D: Society and Space* 24, no. 3 (2006): 427–48.

Hapke, Holly. "Petty Traders, Gender, and Development in a South Indian Fishery." *Economic Geography* 77 (2001): 225–49.

Harbers, Hans, Annemarie Mol, and Alice Stollmeyer. "Food Matters: Arguments for an Ethnography of Daily Care." *Theory Culture Society* 19 (2002): 227–45.

Hayes-Conroy, Alison, and Jessica Hayes-Conroy. "Taking Back Taste: Feminism, Food, and Visceral Politics." *Gender, Place & Culture* 15, no. 5 (2008): 461–73.

Hennion, Antoine. "Pour une pragmatique du gout." *Papiers de recherche du CSI* 001 (2005).

——. "Those Things That Hold Us Together: Taste and Sociology." *Cultural Sociology* 1, no. 1 (2007): 97–114.

"Here Comes The Boss: Loch Fyne tells Tesco to 'f*** off.'" http://www.herecomestheboss. com/blog/hanson-tv/loch-fyne-boss-tells-tesco-to-f-off/ (accessed 8 January 2009).

"Here We Are." www.hereweare-uk.com/aboutus (accessed 4 November 2008).

Hoar, R. "Coming Up Fast: The Man Who Made Oyster His World." *Management Today*. http://www.managementtoday.co.uk/news/411737/Coming-fast-man-made-oyster-world--John-Nobleand-partner-nurtured-specialist-business-precarioustimes-reaping-rich-rewards-spotting-neglectedmarket/?DCMP=ILC-SEARCH (accessed 2 December 2011).

Ingold, Tim. "Evolutionary Models in the Social Sciences." *Cultural Dynamics* 4 (1991): 239–50.

Innis, Harold. *The Cod Fisheries: The History of an International Economy*. New Haven: Yale University Press, 1940.

Jack, I. "Johny Noble." *The Guardian*, 21 February 2002. www.guardian.co.uk/news/2002/feb/21/guardianobituaries.ianjack (accessed 8 January 2009).

Jacobsen, Rowan. *A Geography of Oysters: The Connoisseur's Guide to Oyster Eating in North America*. London: Bloomsbury, 2007.

——. "The Taste of an Oyster." *Chow*. www.chow.com/stories/10713 (accessed 4 November 2008).

Kramer, Kate. "Tracing the Vanishing Woman." *The Drama Review* 36, no. 4 (1992): 158–62.

Latour, Bruno. "How to Talk about the Body? The Normative Dimension of Science Studies." *Body & Society* 10 (2004): 205–29.

——. *Reassembling the Social: An Introduction to Actor-Network Theory.* Oxford: Oxford University Press, 2005.

——. "Why Has Critique Run Out of Steam? From Matters of Fact to Matters of Concern." *Critical Inquiry* 30 (Winter 2004): 225–48.

Loch Fyne Oysters. www.lochfyne.com/About-Us/Philosophy.aspx (accessed 14 October 2008).

Lorimer, Hayden. "Cultural Geography: Non-representational Conditions and Concerns." *Progress in Human Geography* 32, no. 4 (2008): 551–59.

——. "Cultural Geography: The Busyness of Being 'More-Than-Representational.'" *Progress in Human Geography* 29, no. 1 (2005): 83–94.

McQueen, Craig. "The Unlikely Friends Behind Scots Oyster Firm That Became International Hit." *Daily Record,* 14 June 2008. http://www.dailyrecord.co.uk/news/real-life/2008/06/14/the-unlikely-friends-behind-scots-oyster-firm-that-became-international-hit-86908-20606588/ (accessed 2 December 2011).

Méadel, Cécile, and Vololona Rabeharisoa. "Taste as a Form of Adjustment Between Food and Consumers." In R. Coombs et al., eds., *Technology and the Market Demand: Users and Innovation,* 234–53. Cheltenham: Edward Elgar, 2001.

Modleski, Tania. *Feminism Without Women: Culture and Criticism in a "Postfeminist" Age.* New York: Routledge, 1991.

——. "The Rhythms of Reception: Daytime Television and Women's Work." In E. Ann Kaplan, ed., *Regarding Television,* 67–75. Los Angeles: AFI, 1983.

Morris, Meaghan. "A Gadfly Bites Back." *Meanjin* 51, no. 3 (1992): 545–51.

——. *The Pirate's Fiancée: Feminism, Reading, Postmodernism.* Bloomington: Indiana University Press, 1988.

Mount, H. "The Oyster Was Their World." *The Telegraph,* 6 June 2008. http://www.telegraph.co.uk/foodanddrink/3343790/The-oyster-was-their-world.html (accessed 2 December 2011).

Nicholson, Linda, ed. *Feminism/Postmodernism* New York: Routledge, 1990.

O'Neill, Phillip, and Sarah Whatmore. "The Business of Place: Networks of Property, Partnership and Produce." *Geoforum* 31 (2000): 121–36.

Orbach, Susie. *Fat Is a Feminist Issue.* New York: BBS Publishing, 1978.

Palsson, Gisli. "Enskilment at Sea." *Man* 29 (1994): 901–27.

Parasecoli, Fabio. "Postrevolutionary Chowhounds." *Gastronomica* 3, no. 3 (2003): 29–39.

Pratt, Geraldine, and Victoria Rosner. "Introduction: The Global & the Intimate." *WSQ* 34, nos. 1 and 2 (2006).

Prentice, Andrew, and Felicia Webb. "Obesity Amidst Poverty." *International Journal of Epidemiology* 35 (2006): 24–30.

Probyn, Elspeth. "The Anorexic Body." In Arthur Kroker and Marilouise Kroker, eds., *Panic Sex,* 201–12. New York: St Martin's Press, 1987.

——. *Blush: Faces of Shame.* Minneapolis: Minnesota University Press, 2005.

——. *Carnal Appetites: FoodSexIdentities*. London and New York: Routledge, 2000.

——. "Eating for a Living: A Rhizo-ethology of Bodies." In H. Thomas and J. Ahmed, eds., *Cultural Bodies: Ethnography and Theory*, 215–40. Oxford and Boston: Blackwell, 2004.

——. "Feeding the World: Towards a Messy Ethics of Eating." In E. Potter and T. Lewis, eds., *A Critical Introduction to Ethical Consumption*. London: Routledge, 2010.

——. "Glass Selves: Emotions, Subjectivity and the Research Process." In S. Gallagher, ed., *Oxford Handbook of the Self*. Oxford: Oxford University Press, 2011.

——. "How *Do* Children Taste? The Battle of Young Tums and Tongues." In K. Hörschelmann, and R. Colls, eds., *Contested Bodies of Childhood and Youth*. Basingstoke: Palgrave, 2009.

——. *Outside Belongings*. New York and London: Routledge, 1996.

——. *Sexing the Self: Gendered Positions in Cultural Studies*. London and New York: Routledge, 1993.

Radway, Janice. *Reading the Romance: Women, Patriarchy and Popular Fiction*. Chapel Hill: University of North Carolina Press, 1984.

Rich, Adrienne. *Bread, Blood, and Poetry*. New York: Norton, 1985.

Schorr, Naomi. *Reading in Detail: Aesthetics and the Feminine*. London: Methuen Press: 1987.

Sedgwick, Eve K., and Adam Frank. "Shame in the Cybernetic Fold: Reading Silvan Tomkins." In Eve K. Sedgwick and Adam Frank, eds., *Shame and Its Sisters: A Silvan Tomkins Reader*, 1–28. Durham: Duke University Press, 1995.

Silva, Elizabeth. "Gender, Class, Emotional Capital and Consumption in Family Life." In E. Casey and L. Martens, *Gender and Consumption: Domestic Cultures and the Commercialization of Everyday Life*, 141–62. London: Ashgate, 2007.

Simon, Jane. "An Intimate Mode of Looking: Francesca Woodman's Photographs." *Emotion, Society & Space* 3, no. 1 (2010): 28–35.

Skeggs, Bev. "The Making of Class and Gender Through Visualising Moral Subject Formation." *Sociology* 39 (2005): 965–82.

Sloan, D. "The Postmodern Palate: Dining Out in an Individualised Era." In D. Sloan, ed., *Culinary Taste: Consumer Behaviour in the International Restaurant Sector*, 23–42. Oxford: Blackwell, 2005.

St. Martin, Kevin. "The Impact of 'Community' on Fisheries Management in the U.S. Northeast." *Geoforum* 27 (2006): 169–84.

Stengers, Isabelle. "Another Look: Relearning to Laugh." Trans. P. Deutscher. *Hypatia* 15, no. 4 (2000): 41–54.

Tamburri, Mario N., Richard K. Zimmer, and Cheryl Ann Zimmer. "Mechanisms Reconciling Gregarious Larval Settlement with Adult Cannibalism." *Ecological Monographs* 77, no. 2 (May 2007): 255–68.

Thrift, Nigel. *Non-representational Theory: Space, Politics, Affect*. London: Routledge, 2007.

Warde, Alan, Lydia Martens, and Wendy Olsen. "Consumption and the Problem of Variety: Cultural Omnivorousness, Social Distinction, and Dining Out." *Sociology* 33, no. 1 (1999): 105–27.

Whatmore, Sarah. *Hybrid Geographies*. London: Sage, 2002.

——. "Materialist Returns: Practising Cultural Geography in and for a More-Than-Human World." *Cultural Geographies* 13 (2006): 600–9.

Winter, Michael. "Embeddedness, the New Food Economy and Defensive Localism." *Journal of Rural Studies* 19, no. 1 (January 2003): 23–32.

3

JAMAICA KINCAID'S PRACTICAL POLITICS OF THE INTIMATE IN *MY GARDEN (BOOK)*:

Agnese Fidecaro

M y Garden (book): (London: Vintage, 1999) is a collection of essays that Jamaica Kincaid wrote for her gardening column in the *New Yorker* and for magazines such as *Architectural Digest* and *Travel and Leisure*.[1] In this chronicle of Kincaid's gardening practice, she meditates upon the pleasure she derives from it while also dispensing some of the practical information that gardeners are interested in, from her purchases of tools and seeds to her trips to various nurseries and gardens. Thus, the essays fit in and rewrite a heterogeneous tradition of garden writing that includes "numerous classic guides to planting and tending gardens, essays about the joys of gardening, plant catalogues, and discussions about the principles of botany."[2] This tradition is characterized by a small-scale kind of learning and, to the extent that it modestly keeps the scope of their knowledge within the confines of the domestic, has been particularly legitimate for women to pursue. At the same time, it has been implicated in the fashioning of a Western cultivated self who finds fulfillment in the tending of its private property and the metaphorical exploration of its own interiority.

Yet from its empirical beginnings in a series of unskilled experiments with seeds, Kincaid's passion also has a larger scope:

> Then again, it would not be at all false to say that just at that moment I was reading a book and that book (written by the historian William Prescott) happened to be about the conquest of Mexico, or New Spain, as it was then called, and I came upon the flower called marigold and the flower called dahlia and

the flower called zinnia, and after that the garden was to me more than the garden as I used to think of it. After that the garden was also something else.[3]

As she comes across familiar flowers in her readings about colonization, Kincaid starts considering the embeddedness of gardening in a variety of discourses and practices of power: history, botany, the aesthetic; transplantation, hybridization, naming; the eradication of a traditional relation to the environment and the transformation of the West Indian landscape through the development of the plantations. She questions the nature of the enjoyment she experiences in her magnificent garden and underscores the ways in which her own passion for plants is compromised by history: "I have joined the conquering class: who else could afford this garden—a garden in which I grow things that it would be much cheaper to buy at the store? (92)."

While this affiliates Kincaid's texts with public genres such as the journalistic essay or the moral pamphlet, the title of the collection evokes the diary-like genre of the garden book and its careful recording of the most disparate aspects of the everyday. Kincaid radicalizes this conventional domesticity by exploring its continuity with personal memories and emotions connected with trauma, sexuality, and the body. The constitution of an autonomous sphere of the domestic historically had to do with the institution of a rational economy that disciplined the female body and abstracted its materiality into language.[4] By contrast, Kincaid's displacement of the domestic through the intimate redefines the house as a realm of excess, of disorderly materiality and shattering emotions. The intimate thus emerges as that which undermines the closure of the domestic. Kincaid more precisely approaches it as a conflictual, traumatic node that also fissures the global as a sphere with totalizing claims and transforms the conventional discourse of the complementarity of the home and the world. Her always incomplete, open-ended exercises in horticultural creativity and garden writing turn out to be, I will argue, so many instances of a practical politics of the intimate.

The essays subversively develop this politics from within the materialism and superficiality of opulent domesticity. Critics have been rather dismissive of this aspect of Kincaid's garden writing, of her dwelling on the time she has for leisure or on the small worries associated with her hobby. Marie-Hélène Laforest objects to Kincaid's individualism and her reproduction of a Western ideology of achievement (which includes having a large house, a husband, and two kids), both of which entail a "lack of racial solidarity" and a certain rejection of Caribbean culture.[5] Lizabeth Paravisini-Gebert is similarly ambivalent: "A number of [these essays] . . . , particularly those on the subject of seed catalogues and the differences between the services provided by the various nurseries, seem self-indulgently tedious. . . . There is evidence in [them] of a struggle between those themes that are close to Kincaid's heart—genocide, colonialism, cultural erasure, history—and a garden-writing tradition that is in many ways a closed, self-absorbed world."[6] Here the critic constructs an opposition between Kincaid's intimate political concerns and feelings, the value of which she acknowledges, and her recourse to materialist details and an alien form. This gesture, however, reproduces a dichotomy of authentic inside and inauthentic outside that also structures domesticity as an ideological construction.

I would claim that Kincaid's performance of domestic consumerism and discursive self-referentiality is not parasitical to her politics but critical and self-conscious.[7] A central aspect of her postmodern writing, it makes it difficult to construct the intimate as the unmediated source of a more authentic, unambiguous personal politics. If Kincaid looks for ways of turning the house into a site of resistant practice, it certainly is in full awareness of the feminist critiques that define it as a space of alienation, commodified existence, or middle-class privilege.[8] Contrasting with the one-sidedness of such critiques, however, her politics of the intimate involves a complex acknowledgment of ambivalence. Conventional domestic life in her essays harbors powerful emotional and bodily political energies that result from conflictual positionings and traumatic woundings. Domesticity also screens the overwhelming affects connected with them, translating them into surface concerns at the risk of deflating them. Kincaid demonstrates a gardener's awareness that it takes some practical know-how and refusal of dogma to safely unravel their political potential.

The suspicion that Kincaid's taste for plants may ultimately lead her away from politics does not, therefore, take seriously enough the writer's own statements concerning the intrinsic relation of gardens to the larger world that surrounds them. Contrary to Voltaire's Candide, whose decision to cultivate his own garden signified a retreat from public affairs, Kincaid does not dissociate her gardening from an engagement with the world:

> I suspect that the source of my antipathy to Sackville-West and her garden is to be found in her observations of the garden, in the way she manages to be oblivious of the world. For the fact is that the world cannot be left out of the garden. At least, I find it to be so: that is why I regard Nina Simone's autobiography as an essential companion volume to any work of Vita Sackville-West's. There is no mention of the garden in Nina Simone's account of her life, as there

is no mention of the sad weight of the world in Sackville-West's account of her gardening (59).

Kincaid captivates her postcolonial readership when she brings to light some of the details of colonial history, when, for instance, she recounts how the dahlias of her own middle-class garden, far from being indigenous, were introduced in the West as a result of conquest. Similarly, she efficiently disrupts readers' ideas of nature by telling them the story of how these flowers were hybridized and renamed. But the petty details that make up the dimension of the personal will remain suspect of a compromise with middle-class values unless they are immediately translated into a discourse of world-historical significance such as the master narrative of imperialism. The continuing efficacy of a hierarchy that implicitly subordinates the personal to a sphere of the political defined in terms foreign to it thus underlies the critics' unease. According to this partial interpretation of the feminist motto "the personal is political," the personal may offer access to the political, but it cannot redefine it.

I will show that Kincaid's skilled exploration of gardening as a counterpractice of memory effectively resists from inside the consumerist, amnesic reification of the intimate. In their explorations of the personal and mythic longings associated with the garden and the house, the essays particularly perform and rewrite Kincaid's nostalgia for home and a space of her own. But they simultaneously break the closed and circular, potentially regressive temporal structure of nostalgia in favor of an open orientation toward the "to come." Although it has lost its innocence, Kincaid's garden thus remains a utopian space for creativity and personal integrity.

As such, it also rewrites Virginia Woolf's *A Room of One's Own*. It is particularly useful to consider here how *My Garden (book):* ambiguously echoes "In Search of Our Mothers' Gardens," Alice Walker's revision of *A Room of One's Own* from the perspective of black women's need for an empowering artistic tradition.[9] Walker's garden is an alternative utopian, aesthetic space that an underground female tradition hands down to her. This personal territory of spiritual wealth and organic plenty is subtracted from yet remains crucially continuous with the mother's double life of poverty and hard work inside the house and in the fields. In keeping with this sense of continuity, it almost religiously combines the intimate and the communal to positively advocate and celebrate a collectively refound identity.

Kincaid similarly cultivates the seeds of female creativity and resistance in her garden. Yet her recovery of the personal as a resource for the political is much more conflict ridden. Walker's sense of connection with a transfigured tradition of the oppressed is not, indeed, available to Kincaid: however nostalgic for it she may be, she does not inherit her mother's garden but must invent her own from within a prosaic domesticity inextricably bound up with the position of economic privilege and power she has individually achieved. Responding to that sense of separation, her essays offer a disturbing version of the feminist "room of one's own"—one that specifically rewrites and actualizes Woolf's demand for personal, not just aesthetic integrity in the context of the contemporary world. The violence of this world sometimes reaches so deep into our lives that we may be tempted to withdraw from it into the safe space of our personal gardens. Acknowledging this temptation, Kincaid turns their "interiority" inside out and finds new ways of conceiving of the intimate as a political resource we may discover there.

SUBVERSIONS OF PROPERTY

The hybridity and digressiveness of Kincaid's garden writing have been much remarked upon. Subverting the hierarchies that depoliticize gardening (either negatively, by treating it as an insignificant hobby, or positively, by mobilizing the higher category of the aesthetic), she organizes a series of disruptive encounters between high and low literature, autobiography, gossip, art history, commercial seed catalogs and learned gardening treatises, and the aesthetics of the English garden, but also colonial history, the science of horticulture and the history of botany (as a branch of modern science with universalist claims). These subversions of discursive property are continuous with Kincaid's revisions of domestic property, of the domestic as a closed and compartmentalized space. As she deessentializes the Western house and the values associated with it, she turns categories such as the inside and the outside upside-down: "But I do not believe that I know how to live in a house. I grew up outside.... When we started to do things together inside our house, things other than sleeping, it was a sign of some pretension (28)." She also questions the tidiness and order often associated with the inside: "The inside of my house looks like my yard; it is smudged with dirt, it is disorderly for an inside of a house, though it would look wonderful and memorable if it were the outside of the house I grew up in" (31).

The intimate is that personal-historical-bodily dimension that resists and exceeds the opposition between the inside and the outside, the private and the public, the individual and the collective, the body and discourse. The private often depends upon a territorial drawing of boundaries and upon a juridical, thus stable, definition of property. The intimate, by contrast, suggests an inner core that is most immediately experienced as one's own and yet essentially emerges in a conflict over what is proper or as a potentially subversive contestation of property. When Kincaid uses an empty box of sanitary pads to steal, as she claims all gardeners do, some seeds from Monet's garden in Giverny, the intimate is implicated on two levels: it plays a central role in a transgression of property that symbolically reverses the stealing of black women's "seeds" and the appropriation of their fertility in slavery; and it becomes part of a process of transplantation that ensures that the garden is caught in a global circulation of an alternative sort (59).

The intimate is idealized in Western culture, where it is closely bound up with the home and is protected by privacy. Kincaid, however, constructs it as the sphere of the untidy, as the realm of an improper, unruly bodiliness, capable of resisting the various authorities bent on disciplining it. Female sexuality—and those sexual dimensions of the female body, such as menstruation, that cannot be idealized in terms of the sublimating language of the aesthetic—is an aspect of this bodily unruliness. At the same time, Kincaid's prosaic recycling of her box of sanitary pads acknowledges the commodification of the female body and precludes any hasty identification of this metonymically evoked sexual unruliness with "nature."

This unruliness of the female body also finds in the illegitimacy of gossip, the private-public discourse that circulates the intimate for consumption, a fit vehicle. Kincaid is thus happy to find in Gertrude Jekyll's all-too-decent biography a passage explaining how this celebrated British designer of gardens (1843–1932) was not, as a child, allowed to pick up dandelions, which she was told were called "Nasty Things." Wondering, gossip-like, whether Jekyll ever had sex, Kincaid relates the normativity of her discourse about gardens to the sexual repression that may have shaped her supposedly disembodied aesthetic judgment: "What better way to divert attention from herself than to make pronouncements about correctness and beauty in the garden. What a perfect example of making a virtue of your own neuroses!" (68–69).

The episode of the dandelions links naming, plants, and female sexual organs, which are experienced as bad and dangerous. It describes an education in which the disciplining of the female (sexual) body cooperates with the discursive production of value to effect a most intimate subjection to power. It is, as such, akin to the much discussed episode of the daffodil in Kincaid's novel *Lucy*, in which the protagonist's dislike of daffodils is related to her memory of being forced to recite Wordsworth's poem "I Wandered Lonely as a Cloud" at school.[10] In it, the description of a little Antiguan girl's interpellation by colonial education underscores how the disembodied discourse of poetry effects, through the practice of recitation, the internalization of linguistic norms and the acquisition of a "proper" female behavior.[11]

The unruliness of the intimate has the potential to dislocate the domestic, the very sphere that both harbors and domesticates it. It is conceptualized as that which is most "inside," yet its "interiority" cannot be dissociated from the externality of the world-historical processes that construct it. It is not simply "contained" by the domestic and is not the graspable (in)side of a comfortable, totalizing sphere of the global. Instead, it has a reopening power that interrogates from inside all closed and enclosing structures that have to do with (self-)possession or property. Kincaid closely relates the intimate to a material dimension of the body that has traditionally been associated with the abject: "All small events are domestic events, and domestic events are those events that can occur in any area in which it seems quite right to expel saliva. If I were asked to make a definition of domestic space, I would say that domestic space is any space in which anyone might feel comfortable expelling any bodily fluid (23)." The abject, the revulsion with which the self responds to the calling into question of its separation from the world, is not a relevant experience here precisely because the intimate is that "comfortable" sphere in which the rules of proper behavior that repress the materiality of the body and protect the frontier between inside and outside do not prevail.[12]

The garden similarly is a sphere in which the materiality of the body and sexuality reasserts itself. This materiality, for instance, disrupts the cozy, domesticating discourse of gardeners, who apprehend plants in terms of a familiar, stable world: when a catalog describes the "unique carrot shape" of a beet called Formanova, Kincaid does not resist adding that "this beet isn't shaped like a carrot at all, it is shaped like a penis, and I always refer to it that way; I call it the penis-shaped beet (64)." This determination to return a composed discourse to the low sphere of the body may signal an investment in the regenerative function of the carnivalesque and in a temporality that has affinities with processes of growth. This does not involve any nostalgic return to the organic, however: Kincaid disrupts from the start any view of the naturality of the body with her description of her neighbor Chet, who "could breathe properly only while attached to canisters filled with oxygen," who would smoke while tending his vegetables, and who "did not worry about poisonous toxins leaching out of the materials from which his house was built into the soil in which his tomatoes were grown (xi–xii)." Chet's mutilated bodiliness, a hybrid of body and machine, is a living testimony to the violence that world industrial powers exercise on bodies, while his smoking derisorily defies the health authorities that insist on disciplining them. It is from that untenable position, from the excess of that bodiliness, of those unresolved contradictions and painful mutilation, that his gardening paradoxically functions as a resistant assertion of personal integrity—something quite different from making a virtue of one's own neuroses.[13]

THE UNHOMELY

Kincaid's longing for fulfillment and perfection, which is so important to her politics of nostalgia, is asserted from within a similar sense of mutilation. It finds a self-consciously trivialized expression in some of her most blatantly apolitical feelings, such as the "emotional devastation" she experiences when the latest issue of a favorite potato catalog does not live up to her expectations. Gossip, too, is of the utmost value to her, and she defers important intellectual endeavors to enjoy readings such as "the entire six volumes of the Mapp and Lucia saga written by E. F. Benson." But self-parodic trivialization is not to be dissociated from seriousness and poignancy here: "The grimness of winter for this gardener can be eased only by such things. On my night table now is a large stack of books and all of them concern the Atlantic slave trade and how the world in which I live sprang from it. The days will have to grow longer, warmer, and softer before I can pick one of them up (44–45)." This way of ironically giving the most trivial everyday priority over the tragedy of history paradoxically makes a statement about the emotional weight of that history: about the necessity to keep it at a distance while also having it near, on the night table. In that last sense, a politics of the intimate involves entangled ambivalent feelings that are perhaps less easily objectifiable than a knowledge of the political history of the dahlia.

The unhomely is both summoned up in its most personal implications and kept at bay here. Homi Bhabha has analyzed the disturbing potential of this "paradigmatic postcolonial experience," which describes the moment when "the intimate recesses of the domestic space become sites for history's most intricate invasions."[14] Any utopia of the complementarity between the garden and the world must accommodate the traumatic—in Bhabha's words, the uncontained "anguish of cultural displacement and diasporic movement."[15] Yet Kincaid's version of the unhomely cannot be reduced to the traumatic as a generic term describing the dispossession endured in the face of conquest or the shattering of language by violence. Trauma does play a decisive role in the unhomely, which "relates the traumatic ambivalences of a personal, psychic history to the wider disjunctions of political existence."[16] But if the unhomely is to be approached as a moment of resistance, we must further explore its ambivalence to determine how the destructiveness of trauma may be articulated, through the (sexual, carnivalesque) body, with sources of personal and collective integrity.

The attention Kincaid pays to the sphere of the body—to its sexuality and the traumatic memory it encloses—precisely allows her to define on her own terms the conditions of her participation in the exchanges that make up the so-called global world. As the partly autobiographical episode of the daffodils makes clear, her violent interpellation by the impersonal structures of imperialism would entail her psychic annihilation—a prerequisite for her cultural assimilation or cooptation: "The night after I had recited the poem, I dreamt, continuously it seemed, that I was being chased down a narrow cobbled street by bunches and bunches of those same daffodils that I had vowed to forget, and when finally I fell down from exhaustion they all piled on top of me, until I was buried deep underneath them and was never seen again."[17] Yet it is her turning of such a destructive interpellation back upon itself—her mobilization of trauma as that which is most personal to her and yet most historical—that allows for a mode of resistance that is coterminous with the deployment of an unrestrained creativity. Resistance, in other terms, does not arise from any preserved territory of the self, but from an expropriated intimacy that is inextricably implicated in the historical processes that have constructed today's world and go on shaping it. The body here is the site of an unmediated experience of power: not as a sphere of innocence that could be redeemed from devastation, but as the keeper of a record that has yet to be written because it partakes of both language/discourse and what resists it.

A COUNTERPRACTICE OF MEMORY

As we have seen with her desacralizing treatment of Gertrude Jekyll, Kincaid's questioning of authority and value forms a significant aspect of *My Garden (book):*. This contestation comes from a practical perspective that refuses preconceived norms and aesthetic hierarchies. An authoritative statement concerning the enjoyment to be derived from "green grass kept finely shorn" is thus undermined by her observation that no one who actually has to cut the grass would make such a comment: the discourse of the aesthetic (which here naturalizes green turf, a significant aspect of England's national self-image) is criticized for its bracketing out of the dimension of practice (83), and one cannot help thinking of Bourdieu, who denounces the shortcomings of theory on similar grounds.[18]

Kincaid's habit of providing the kind of trivial detail that only gardeners will want to know, such as where she bought her set of hand tools, also valorizes what is irreducibly particular, what resists being integrated into a prescriptive, general script. To her, all discourse concerning a gardener's good taste or "inherent feeling for design and good color combinations" is "bullying," and she writes ironically about Jekyll's comments on "between" plants, which have the subordinated function of enhancing their neighbors (58–59). In response to Jekyll's view that the capacity "to apprehend gardening as a fine art" hinges on the ability to make the difference between a garden and a collection, Kincaid asks: "But what if all the flowers I love and want very much to grow are, when seen together, all wrong, all jarring and displeasing?"(66). This is no rhetorical question: Kincaid's practical ethics of gardening, which leads her to think of each flower "by itself, isolated, disregarding how it might fit into the garden as a whole" (67), entails a release of control over the overall outcome: "I don't know how this will look. It may not only violate established rules, but also not please me in the end at all" (67).

Kincaid here mimes the voice of the experienced amateur who has developed a certain idiosyncratic style of performing the most functional tasks. Her relation to the garden, far from being characterized by the detachment of the aesthete, is thick with personal reminiscences. The traditional links of garden writing to autobiography,[19] a sign of the implication of gardening in the fashioning of the Western private, cultivated self I mentioned in my introduction to this essay, acquire a new significance here. The radical potential of a practice of writing that would allow the irreducibly personal to assert itself as a counterpoint to the most specialized discourses and areas of knowledge is retrieved.

When Kincaid starts her first garden and her flower beds acquire the most unlikely shapes, it is only after the fact and as a result of practice that she realizes what she has been doing:

> When it dawned on me that the garden I was making (and am still making and will always be making) resembled a map of the Caribbean and the sea that surrounds it, I did not tell this to the gardeners who had asked me to explain the thing I was doing, or to explain what I was trying to do; I only marveled at the way the garden is for me an exercise in memory, a way of remembering my own immediate past, a way of getting to a past that is my own (the Caribbean sea)

and the past as it is indirectly related to me (the conquest of Mexico and its surroundings) (xiv).

Here Kincaid describes the workings of a bodily memory that does not rely upon any conscious processes. The past becomes the product of a practice that creates and invents, in an ongoing process, what it remembers. The result is a personal "map" that counteracts the objectifying function of cartography, an imperial science that abstracts the land from people's embodied experience of it. In it, the dimension of bodily practice and the dimension of discourse are inextricably linked.

Far from functioning as a discourse of the mere subordination and domestication of nature and the body, gardening mobilizes those dimensions of the self that resist objectification. Thus Kincaid's intimate and caring, but also restless, rather than aesthetically detached and poised, relation to plants allows her to retrieve a moment of awakening sexual awareness that had not been fully experienced at the time. She belatedly understands, as she is kneeling over a portulaca, "fretting about its health," why she likes the flower so much: she remembers that as a child, she used to be left alone in front of two banks carpeted with this flower while the woman who was looking after her would go inside the house to have sex with her boyfriend. The memory of this episode of deferred sexual initiation (it is only "some time later" that she understands what is going on) has thus inscribed itself in the restless body of the little girl dancing up and down around the flowers, rehearsing/coming to terms with her own exclusion by playing at being a stranger to the place.[20]

EXPERIENCING LOSS: THE POLITICS OF NOSTALGIA

Gardening as an ongoing practical "exercise in memory" allows Kincaid to explore her own nostalgia. In both passages—the fretting about the portulaca and the garden as a map of the Caribbean—the body seems to be inscribed with a memory of painful loss that only belatedly asserts itself. As a map of the Caribbean, the garden is an expression of Kincaid's nostalgia for the West Indies, while the portulaca would perhaps not yield so powerful a memory if it was not connected with her separation from the mother: the episode is about sexual awareness, but what gets inscribed in the body, becoming inextricably implicated into sex, might also be the memory of the mother handing her over to a babysitter who then "abandons" her in her turn.[21] In that sense, Kincaid's emphasis on an affective relation to the garden, her foregrounding of the material, embodied dimensions of culture, and the priority she gives to the spontaneity of practice over theoretical understanding and planning all need to be related to the role bodily memory plays in the processing of trauma.

Trauma is triggered by a historical violence that reaches far into the self. It is connected with a pain that has not reached closure, that still deploys its incapacitating effects. That pain still demands to be rehearsed and returned to because something, in the event that caused it, eludes the possibility of once and for all naming it and putting it at a distance:[22] "What to call the thing that happened to me and all that look like me? Should I call it history? And if so, what should history mean to someone who looks like me? Should it be an idea; should it be an open wound, each breath I take in and expel healing and opening the wound again, over and over, or is it a long moment that begins anew each day since 1492 (124–25)?" Here trauma appears to partake of both discourse (history, an idea) and body (an open wound). "The thing that happened to me" is an illocalized event that only discontinuously reveals itself. Kincaid's breathing—perhaps as painful as Chet's, the man with the canisters of oxygen—crosses and recrosses the boundary between the inside and the outside of her body, between intelligible idea and unintelligible pain. It articulates the inextricability of violence and healing as the two moments of a bodily rhythm. Contrasting sharply with the linear temporality of history, this specific scansion is not unrelated to the rhythm of gardening as a practice that experiences the materiality of the body, of plants, or even of books in its unresolved connection with and separation from the intelligibility of discourse and reading. Through gardening, Kincaid incorporates pain, loss, and their overcoming into the ongoing rhythms of practice, rehearsing them without reaching any definitive closure.

The belated functioning of traumatic memory disrupts the logic of the once and for all, a logic that leaves no room for the unexpected and for what resists the imposition of a rational order. Kincaid thus contrasts the idea of an estate being a part of the national heritage with

> something crucial [that] had been lost over time: the sense of the place not as some sort of national park but as a piece of land a man arranged out of who knows what psychological impulses. . . . I was in a country whose inhabitants . . . do not know how to live in the present and cannot imagine living in the future, they can live only in the past, because it, the past, has a clear outcome, a winning outcome. A subdued nature is part of this worldview in which everything looks beautiful (82–83).

Just as the garden taken as a fine art valorizes a certain aesthetic closure (we can also think of the etymological link of the garden with an enclosure here), the garden as national monument turns history into something that is over, that has already happened. It leaves no room for the event or for the "impulses" that may determine, beyond any planning, the arrangements one makes. Plants disrupt such closure, just as practice always encounters the resistance that the otherness of the (natural) world offers to our schemes. Kincaid does not, therefore, only miss the garden/the Caribbean in the nostalgic sense of the term: she also misses it in the sense of a deliberate staging of failure rather than mastery. When she realizes that the untidy sprouts that are spoiling her lawn are peonies planted by the previous owner, the discovery entails a form of dispossession, the realization that places have a kind of stubborn memory that cannot simply be suppressed by the new owner. This is the opposite of the erasure that the colonial appropriation and reshaping of her native landscape involved (100–101, 119). Kincaid's own relation to her house is mediated by the emotions of a young man who has grown up in it and by his most intimate memories. Plants again are channels for the uncommunicable, which is conveyed across the dispossession the young man is now experiencing: "When he sees the trees and when he speaks of the trees, he is speaking of things that he is perhaps conscious of, perhaps not, but that are not being communicated clearly, and should not be communicated clearly" (21).

Forced closure is, I would argue, a problematic way of dealing with the pain of nostalgia, with the longing for what has been lost. Nostalgia looks back to what was and longs to restore it in its completeness, to repossess it, to heal it. An implicit awareness of the complicity between nostalgia and the desire for possession is expressed when Kincaid relates the Europeans' transplantation of the dahlia to their "longing" for it: "Who first saw it and longed for it so deeply that it was removed from the place where it had always been, and transformed (hybridized) and renamed?" (88). A specific form of longing, the desire for permanence and perfection (in the sense of something that has been perfected, that is finished) underlies colonial knowledge and the aesthetics that corresponds to it. The fruits and flowers of blown glass exhibited at the Museum of Natural History at Harvard trigger the following comment:

> The creation of these simulacra is also an almost defiant assertion of will: it is man vying with nature herself. To see these things is to be reminded of how barefaced the notions of captivity and control used to be, because the very fabrication of these objects, in their perfection (no decay or blemish in nature is

ever so appealing) and in the nature of the material from which they are made, attests to a will that must have felt itself impervious to submission. How permanent everything must feel when the world is going your way! (56–57).

Against that aesthetic permanence, which associates the passion for possession to the production of an incorruptible materiality, Kincaid asserts the value of a different, less transcendent materiality. Taking her books with her in the garden, she allows the materiality of gardening, of mud, bad weather, and the physicality of the body, to contaminate them: "I read my books, but I also use them; that is, sometimes the reading is almost a physical act" (57). She also expresses the deferral of satisfaction in gardening in terms that are suggestive of utopian expectations rather than regret: "An integral part of a gardener's personality—indeed, a substantial amount of a gardener's world—is made up of the sentiment expressed by the two words 'To Come'" (61).

A myth of lost enclosure, self-containment, and perfection, that of the Garden of Eden, gives a nostalgic turn to both the colonial relation to exotic lands and the Western practice of gardening. As John Prest has demonstrated, the history of the botanical garden, which involved the appropriation and transplantation of numerous species of plants, was not just a scientific venture, but was underlain by the mythic desire to recreate Eden.[23] Kincaid presents Carol Linnaeus, naming plants in a glasshouse and inventing modern plant nomenclature, as being "Adamlike (112, 124)." Her own concern for naming trauma, which I discussed earlier, makes her Eve-like as she critically questions, in the name of a pain incommensurable with any "universal standard," the desire and intimate motivations underlying the production of such nomenclatures.[24]

The Eden of botanical gardens also is a perversion: as part of a quest for rationality that she questions, botanists "emptied worlds of their names; they emptied the worlds of things animal, vegetable, and mineral of their names and replaced these names with names pleasing to them" (120).[25] Kincaid positively describes the botanical garden of her childhood, where she used to sit with her father, as "an enormous expanse of land, Edenic, in my memory" (106). Yet she also associates Eden's perfection with the patriarchal authority of the first Gardener. Acknowledging her longing for "the garden I have in my mind," she asserts that she will never have it, yet makes desire—longing as Eve's original sin?—the principle presiding over its continued creation: "A garden, no matter how good it is, must never completely satisfy. The world as we know it, after all, began in a very good garden, a completely satisfying garden—Paradise—but after a while the owner and the occupants wanted more" (169–70).

The ambivalent feelings of nostalgia and longing also play a central role in Kincaid's own complex relation to the domestic. The acquisition of her house is presented as a result of longing and reflects a desire for possession: "I longed to live in this house, I wanted to live in this house" (18). She does value the stability of settling down, which she relates to mothering ("for that sort of settling down is an external metaphor for something that should be done inside, a restfulness, so that you can concentrate on that other business, living, bringing up a child" [18]). Yet she also takes her distance from an earlier fascination for the domestic, which was a projection of her nostalgia for a lost home: "When I was young and living away from my family, my life was almost completely empty of domestic routine, and so I made a fetish of the way ordinary people in families lived inside their homes. I read women's magazines obsessively and would often cook entire meals (involving meat in tins and frozen vegetables) from the recipes I found in them" (64). On the one hand is the temptation of indulging a nostalgia that fetishizes home, suggesting that you can hope to possess it by buying the magazines that provide virtual, consumerist versions of it; on the other hand is Kincaid's less regressive investment in the domestic, which is linked to the practice of gardening she explores.

Kincaid does accept some of the mystique associated with the idea of "home," yet she rewrites it as an ethical Utopia without any normative content, one that is entirely hers to invent, since the recipe for it has been lost, leaving only longing in its place:

> Oh, how we wish that someone . . . had given us a recipe for how to make a house a home, a home being the place in which the mystical way of maneuvering through the world in an ethical way, a way universally understood to be honorable and universally understood to be ecstatic and universally understood to be the way we would all want it to be, carefully balanced between our own needs and the needs of other people, people we do not know and may never like and can never like, but people all the same who must be considered with the utmost seriousness, the same seriousness with which we consider our own lives (33).

The ethical praxis described here is "ecstatic": it is not a withdrawal inside in self-possession, but a going out of oneself into the world. Kincaid's garden has thus little to do with the nostalgia for a self-contained, integrated sphere of self-possession and being. On the contrary, in her final essay, titled "The Garden in Eden," the reader discovers that its enclosure has been turned inside out and has become the world: "Is this Eden, that thing that was banished, turned out into the world as I have come to know it—the world of discarding only to reclaim, of rejecting and then claiming again, the world of such longing that its end (death) is a relief?" (171). The garden is the world, inhabited by the traumatic scansions of loss and healing, of possession and dispossession. Kincaid has, at the point of coming to think of this, left the confinement of her own garden and is restlessly walking across the Chinese landscape: "I was in the wild, the garden had become the wild and I was in it (even though all the time I was really in China)" (175). The housebound mother and the cosmopolitan globe-trotter, who travels to China in search of seeds, have merged into a hybrid figure, as further suggested by Kincaid's *Among Flowers*, a volume that crucially makes travel writing the counterpart of her garden book. Refusing to be either locked out or locked in, questioning the opposition between the garden and the world, Kincaid may well send us back in an unexpected way to the restless, improvisational Virginia Woolf of *A Room of One's Own* and offer a newly subversive, postcolonial version of her explorations of space and quests for knowledge, poetic credos, challenges to authority, and coming to terms with the wounds of history.

A POSTCOLONIAL "ROOM OF ONE'S OWN"

While gardening possesses its own irreducible goals, practical structures, and temporal constraints, it also functions as a metaphor for writing. Kincaid's essays deal with the politics and aesthetics of planting and weeding, but also the politics and poetics of creativity. In spite of its subversion of existing discourses on gardening and its aesthetics, *My Garden (book):* still approaches the garden as a privileged space for creativity and experimentation. Kincaid's postmodern awareness of its complex, shifting discursive makeup ("the garden for me is so bound up with words about the garden, with words themselves, that any set idea of the garden, any set picture, is a provocation to me" [xiii]) does not dissolve its utopian potential. Kincaid's garden thus is a room of one's own, that personal space distinct from the family in which women may create and grow.

Kincaid's garden was started on Mother's Day, when she received some seeds and tools from her husband. This would tend to make her gardening continuous with her roles as a wife and mother, an association reinforced by the fact that the essays are dedicated to her children. However, the garden is connected with intellectual and political concerns as well as personal memories and emotions that have little to do with being a mother. We have seen that it takes Kincaid around the world, in a kind of critical rehearsal of the trips undertaken by the botanists of empire. The book's dedication redefines domestic roles in two opposite ways: on the one hand, it refers to the emotional closeness between mother and children in strong, intimate terms that mime, yet also exceed, the conventional, socially controlled expression of motherly love; on the other hand, it presents the garden as that which introduces some spacing in this intense relationship: "With blind, instinctive, and confused love,/for Annie & for Harold/who from time to time are furiously certain that the only thing standing between them and a perfect union with their mother is the garden, and from time to time, they are correct." Although a strong tension arises between the two poles that the dedication establishes, they both involve overwhelming passions that destabilize from within the more controlled, socially established realms of which they partake. Kincaid's personal space does not fit, then, any neat opposition between the alienating demands of the family and the autonomy of a separate space that would allow women to repossess themselves.

Kincaid's experiments with gardening certainly redefine the room of one's own in terms of a postcolonial context shaped by a history of diaspora and transplantation rather than rootedness, by dispossession rather than "owning." But the essays also insist on the materiality of that space and on its connection to the body. The feminist room of one's own entails something more, then, than the peacefulness, privacy, and concentration necessary for writing to occur: the intimate as a politically significant, disruptive dimension. Kincaid's room may have to be wrested from the demands of domestic life, but it is deeply in touch with the subversive sphere of the intimate that lies at its core, and it is from that connection and that sense of the relationality of spaces that it interpellates the contemporary world.

The concern for women's creativity and aesthetic integrity in *A Room of One's Own* is related to deep anxieties concerning historical and personal trauma. Woolf resolves them with the myth of the androgynous mind, which is supposed to heal the fissures of the world she is living in. Kincaid's open, hybrid aesthetics rethinks Woolf's demand for integrity in terms of the traumas of the contemporary world. Her ways of missing the garden function as a complex response to the traumatic, uprooting dimensions of colonial history. The history of transplantation and loss she deals with defeats, however, any project of a recovered cultural plenitude, self-containment, rootedness, or original completeness. Her ambivalent, conflict-ridden manifesto thus does not promise any forced synthesis, nor are its contradictions resolved in any way. Yet it does propose new ways of thinking together, again and again, the global and the intimate. The room of one's own is reconceived as a space in the making, the space of our intimacies, that we go on cultivating, encountering trauma and recovering sources of material resistance—always missing the garden, always longing for it.

NOTES

1. I would like to thank Susie O'Brien, Maria Josefina Saldaña-Portillo, and the other anonymous reviewers for their most constructive comments and suggestions.
2. Lizbeth Paravisini-Gebert, *Jamaica Kincaid*, 45.
3. Kincaid, *My Garden (book):*, xiii.
4. See Nancy Armstrong's genealogy of the domestic (Armstrong, "The Rise of the Domestic Woman").
5. Marie-Hélène Laforest, *Diasporic Encounters*, 217–22.
6. Paravisini-Gebert, *Jamaica Kincaid*, 40.
7. The politics of consumerism is clearly an issue when one writes for the *New Yorker*. For a critical view of Kincaid's "entangled" relation to the magazine, see Anne Collett, "Gardening in the Tropics."
8. For a deconstructive discussion of feminist critiques of the domestic, see, e.g., Rachel Bowlby, "Domestication."
9. Alice Walker, "In Search of Our Mothers' Gardens."
10. Jamaica Kincaid, *Lucy*, 17–19.
11. For an analysis of this "rite of colonial obedience" and a discussion of the erasure of the female body in it, see Helen Tiffin, "'Replanted in This Arboreal Place.'" Tiffin seems to posit the existence of an autonomous sphere of the body that could be retrieved outside of discourse. My essay is closer to O'Brien and her insistence on the reciprocal implication of discourse and the physical world in Kincaid. For other discussions of the daffodil episode, see Alison Donnell, "Dreaming of Daffodils"; Moira Ferguson, *Jamaica Kincaid*, 111–13; Jana Evans Braziel, "Daffodils, Rhizomes, Migrations."
12. On the abject and the question of boundaries, see Julia Kristeva, *Powers of Horror*, chapter 1.
13. For a discussion of Kincaid's critical relation to environmentalism, see Susie O'Brien, "The Garden and the World." In the present essay I develop, from a different perspective, the implications of two of O'Brien's points: the idea that Kincaid's garden writing involves an "aesthetic pleasure that does not fulfill desire" and the idea that culture and nature in Kincaid are both incommensurable and inextricably linked (179).
14. Homi Bhabha, "The World and the Home," 446, 445.
15. Ibid., 446.
16. Ibid., 448.
17. Kincaid, *Lucy*, 18.
18. Pierre Bourdieu, *Outline of a Theory of Practice*.

19. See Paravisini-Gebert, *Jamaica Kincaid*, 39.

20. Kincaid, *My Garden (book):*, 67–68.

21. On the politics and poetics of home and displacement in Kincaid, see Antonia MacDonald-Smythe, *Making Homes in the West/Indies*. On the parallels between the relation to the mother and the relation to the mother country in Kincaid, see Ferguson, *Jamaica Kincaid*.

22. On the definition of trauma and the role belatedness plays in its structure, see Judith Lewis Herman, *Trauma and Recovery*; Shoshana Felman and Dori Laub, *Testimony*; Cathy Caruth, *Unclaimed Experience*.

23. John Prest, *The Garden of Eden*.

24. Kincaid, *My Garden (book):*, 124–25. For a critique of the politics of naming connected with botany and the botanical garden, see Collett, "Gardening in the Tropics."

25. Extending John Prest's discussion, Tiffin describes the Caribbean "paradise" as itself "parodic, ironic, and tragic" (Tiffin, "Replanted in This Arboreal Place," 152).

BIBLIOGRAPHY

Armstrong, Nancy. "The Rise of the Domestic Woman." In Nancy Armstrong and Leonard Tennenhouse, eds., *The Ideology of Conduct: Essays on Literature and the History of Sexuality*, 96–141. New York: Methuen, 1987.

Bhabha, Homi. "The World and the Home." In Anne McClintock, Aamir Mufti, and Ella Shohat, eds., *Dangerous Liaisons: Gender, Nation, and Postcolonial Perspectives*, 445–55. Minneapolis: University of Minnesota Press, 1997.

Bourdieu, Pierre. *Outline of a Theory of Practice*, trans. Richard Nice. Cambridge: Cambridge University Press, 1987.

Bowlby, Rachel. "Domestication." In Diane Elam and Robyn Wiegman, eds., *Feminism Beside Itself*, 71–91. New York: Routledge, 1995.

Braziel, Jana Evans. "Daffodils, Rhizomes, Migrations: Narrative Coming of Age in the Diasporic Writings of Edwidge Danticat and Jamaica Kincaid." *Meridians* 3, no. 2 (2003): 110–31.

Caruth, Cathy. *Unclaimed Experience: Trauma, Narrative, and History*. Baltimore: Johns Hopkins University Press, 1996.

Collett, Anne. "Gardening in the Tropics: A Horticultural Guide to Caribbean Politics and Poetics, with Special Reference to the Poetry of Olive Senior." *SPAN* 46 (1998): 87–103.

——. "A Snake in the Garden of the *New Yorker*? An Analysis of the Disruptive Function of Jamaica Kincaid's Gardening Column." In Gerhard Stilz, ed., *Missions of Interdependence: A Literary Directory*, 95–106. Amsterdam: Rodopi, 2002.

Donnell, Alison. "Dreaming of Daffodils: Cultural Resistance in the Narratives of Theory." *Kunapipi* 14, no. 2 (1992): 45–52.

Felman, Shoshana, and Dori Laub. *Testimony: Crises of Witnessing in Literature, Psychoanalysis, and History*. New York: Routledge, 1992.

Ferguson, Moira. *Jamaica Kincaid: Where the Land Meets the Body*. Charlottesville: University of Virginia Press, 1994.

Herman, Judith Lewis. *Trauma and Recovery*. London: Basic Books, 1992.

Kincaid, Jamaica. *Among Flowers: A Walk in the Himalaya*. Washington DC: National Geographic, 2005.

——. *Lucy*. New York: Farrar, Straus and Giroux, 1990.

——. *My Garden (book):*. London: Vintage, 2000.

Kristeva, Julia. *Powers of Horror: An Essay on Abjection*, trans. Leon S. Roudiez. New York: Columbia University Press, 1982.

Laforest, Marie-Hélène. *Diasporic Encounters: Remapping the Caribbean*. Naples: Liguori, 2000.

MacDonald-Smythe, Antonia. *Making Homes in the West/Indies: Constructions of Subjectivity in the Writings of Michelle Cliff and Jamaica Kincaid*. New York: Garland, 2001.

O'Brien, Susie. "The Garden and the World: Jamaica Kincaid and the Cultural Borders of Ecocriticism." *Mosaic* 35, no. 2 (2002): 168–83.

Paravisini-Gebert, Lizabeth. *Jamaica Kincaid: A Critical Companion*. Westport, CT: Greenwood Press, 1999.

Prest, John. *The Garden of Eden: The Botanic Garden and the Re-creation of Paradise*. New Haven: Yale University Press, 1981.

Tiffin, Helen. "'Replanted in This Arboreal Place': Gardens and Flowers in Contemporary Caribbean Writing." In Heinz Antor and Klaus Stierstorfer, eds., *English Literatures in International Contexts*, 149–63. Heidelberg: Carl Winter Universitätsverlag, 2000.

Walker, Alice. "In Search of Our Mothers' Gardens." In her *In Search of Our Mothers' Gardens: Womanist Prose*, 231–43. San Diego: Harcourt Brace Jovanovich, 1983.

Woolf, Virginia. *A Room of One's Own*. London: Grafton, 1977.

4

WIDENING CIRCLES

Rachel Adams

The resident who stitched me up looked improbably young. I watched his head bobbing between my thighs. Everything was numb enough to block the pain, but still I could feel him handling the most private recesses of my body. I can't remember much about his face, but I know he wore a surgical cap made of colorful print fabric and that his hair was buzzed close to his head, military style.

He finished what he was doing and stepped back to survey his work. Concluding that it was good enough, he told me that the stitches would fall out within a few weeks and left the room.

He must have known, but he didn't show it. He didn't say anything about it to me, and I never saw him again. I guess he thought that his job was done once the baby was delivered. Breaking the news would be left to someone else.

In the stillness that followed, I felt a tremendous sense of calm. A nurse kept moving my new baby to different positions, trying to get him to latch on to my breast. His mouth opened and closed weakly. No sound came out. The room was completely quiet and filled with watery winter sunlight.

A pediatrician from Neonatal Intensive Care arrived. She introduced herself to my husband Jon and me, and told us she was going to examine our baby. She spread him on a heated table and turned her back. A few minutes later, she wrapped him up and handed him to me.

"I was called here because your baby has features consistent with Down syndrome," she said. "He's pink and he looks healthy, but we're going to have to run some tests to make sure his heart is functioning properly."

The world should have stopped.

"I know this is a lot to take in. Do you have any questions for me?"

Of course I had questions. *How could this be happening? Couldn't you give me just a few more minutes to believe that my baby is perfect? Why did you have to snatch away my fantasy that anything is possible so quickly?*

I shook my head.

"I'll be back later when you've had time to digest this. We can talk more then," she said kindly. She placed our baby in a wheeled cart and pushed him out of the room.

After she left, the room was still quiet and sunny. I still felt calm. I was enveloped in a numbing blanket of hormones. But somewhere beneath the surface, I knew that fear, grief, and rage were roiling.

Jon and I sat in silence. There was nothing to say. I couldn't cry, thanks to the hormones. But I had already begun to mourn the death of the son I thought I was going to have, and the family I imagined we would be. I remember feeling utterly alone, my entire world compressed into the hospital bed. Who among our accomplished, overeducated friends, with their perfectly healthy, perfectly normal children, would understand our grief? Who else, in this city filled with strivers and go-getters and overachievers, would be reckless enough to forgo an amniocentesis at the age of thirty-eight? Where would we find a community to welcome this fragile, unexpected baby?

My mind raced back over the choices that had brought us to this point. When the time came to make decisions about prenatal testing, ours were on the unorthodox side. For most ambitious, professional women of my age, amniocentesis is a requisite stage of pregnancy, a reassuring guarantee of good health and normalcy. Having devoted a good many years to the study of disability, I had different ideas about the knowledge those tests can yield. I knew that no test could guarantee a perfect baby, or even a perfectly healthy one. And I knew that whatever information the genetic analysis revealed, it couldn't predict the person my child would become. When the results of blood work and ultrasound revealed almost no chance that our fetus had Down syndrome or any other genetic disorder, we declined an amnio. We very much wanted this baby, the tests suggested that he was very likely to be healthy, and we didn't want to risk losing the pregnancy. Besides, we reasoned, given all of the things that could go wrong with a child, Down syndrome wasn't even near the top of our list of undesirables.

Things looked very different from the NICU, as I watched my new baby through the plastic walls of his incubator, where he lay in tangle of tubes and wires. It was one thing to profess my appreciation for difference, quite another to contemplate bringing it home and making it a part of my life. My research had made me painfully aware of the discrimination faced by people with disabilities in the United States. I knew that my baby—with his extra twenty-first chromosome—had been born into a world that has little tolerance for his kind of imperfection. Later, I would learn that as many as 90 percent of all women choose to abort a fetus diagnosed with Down syndrome. Henry was less than an hour old, and already I was worried about where he would go to school, if he would ever have friends, and whether I was capable of giving him the care he needed.

The hospital offered no answers. Over the next week, Henry was examined by countless doctors. All were attentive and thorough. They answered our questions and explained their concerns about his heart, his muscle weakness, and his problems with feeding. We discussed the treatment he might need over the next weeks and months. Our conversations were always about symptoms. Nobody asked what we were feeling or how we were coping with the unanticipated prospect of raising a child with a disability. Nobody offered to direct us toward people who had shared our experiences. When I asked the hospital's social worker if there was a support group that I could contact, she seemed at a loss. It was as if Henry were the only baby with Down syndrome ever to be born on her watch.

I spent that first long night after Henry's birth alone at the hospital. Lying awake at home, Jon reread Michael Bérubé's *Life As We Know It*, about the first three years in the life of his son Jamie, who has Down syndrome. The crazy thing is that I know Michael Bérubé, who teaches American literature at Penn State University. Our paths had crossed a number of times at conferences, and I once invited him to give a talk for a lecture series I run at Columbia. *Life as We Know It* was the first book I ever reviewed, back when I was a graduate student who never imagined having children of my own. My interest in the field of Disability Studies was purely academic, and I thought Michael Bérubé got it better than just about anyone else.

As soon as I was back on e-mail, and before I told any of our friends in New York about Henry, I sent a message to Michael:

Dear Michael,

My husband, Jon, and I have been rereading your beautiful book, which suddenly has a new meaning for us. Last Monday morning our second son, Henry, was born. Twenty minutes later we were told he had "features consistent with Down syndrome." And over the next few days this careful phrasing evolved into a more definitive diagnosis, although we still haven't received genetic confirmation. I think you can understand the shock of such news and the effort to manage the jarring disjunction between the particularity of your own beautiful baby and the category that others are already using to define him. In contrast to many of the books and articles we've already read, we've found your stories about life with Jamie to be so heartening and your perspective on what it means to have a disabled child so thoughtful and sensitive. I hope that we can find the kind of wonder and inspiration in Henry's arrival into our family that you have found in Jamie, and that he will teach us to see the world in new ways, as Jamie clearly has done for you. At any rate, you have been in my thoughts so much in the last days that I wanted to write and share our news with you.

Very sincerely yours,

Rachel

I will always treasure the message he sent in response.

Dear Rachel,

Wow. My goodness. Well, let's just say I wasn't expecting this e-mail. But first things first—a hearty welcome to Henry! May he always find the world as warm as his parents' arms. And congratulations to his wonderful parents! Of course, I know the parental shock well, though back in 1991 my information was so outdated that I thought Jamie's life expectancy would be about twenty-one or so—and now, of course, I couldn't bear the thought of having him with us for only five more years.

You're probably inundated with books and articles and advice, and worries about what the next couple of years—or next couple of decades—will bring. I don't want to wish away those worries; they're real worries, as you know, and the most immediate thing is Henry's health. I hope he's thriving and happy—and that all your doctors and medical personnel understand the concept of "healthy baby with Down syndrome." But over the next few months or so, you'll probably find that the truism is true: babies with Down syndrome are babies first. Henry may reach those grasping-crawling-talking-walking developmental milestones at his own pace, but he'll surely take the same delight in music and play and stimulation that his older brother does, and he'll surely be every bit as beautiful as he is now.

Most of all, I share your hope that you find in Henry's arrival the kind of joy we've experienced with Jamie. About our chaotic human lives one can make no promises, but here's the latest from him: just last month, Jamie proposed that he and I go to New York, because (as he pointed out) we haven't visited the city all year. And because he's now sixteen, I decided to stretch him a little. We dined Friday night at the Plataforma Churrascuria on 49th, played games at ESPN Zone til midnight, slept til 9, got up, hit the Metropolitan Museum at 11 on Saturday (he's on an early modern-baroque binge, and bought a coffee-table book of Caravaggio's greatest hits), went to the Lion King, grabbed a pastrami sandwich for dinner and a 2 train to BAM for some surrealist Japanese dance. Sunday morning it was salsa at Carnegie Hall. He was kinda bored from time to time at the surrealist dance, but then, plenty of other people were too. (Unlike the people who got up and left, Jamie just flipped through his Caravaggio book during the dead spots.) His only disappointment that evening was that we didn't take the Q train over the Manhattan Bridge. The rest of the weekend he spent rubbing his hands together in glee. I know this may seem a bit impertinent when you're dealing with a newborn, but Rachel, when Jamie was a little neonate, nothing made me feel more reassured than the testimonies of older parents with older children, who could tell me from experience that life—in its somewhat altered form— would indeed go on. So I hope this helps. What I'm saying is that you never know. The next few years might be a strain at times, but you're in the greatest city in the world at the best time in history for children with Down syndrome. I know you're inundated with advice, but entre nous, I'd say start the salsa concerts early.

Seriously, our hearts and thoughts are with you and Jon. We know very well how weird and disorienting this time can be. (May I forward your letter to Janet?) If you need to get in touch with us for any reason, please feel free to call.

And best wishes to all your family from all of ours,

Michael

Michael understood the importance—and difficulty—of separating our baby from his diagnosis. The story he told about Jamie was far from impertinent. What it said to us was: your life is not over. You may not believe it now, but you will be happy again. Like other children, your child will bring you joy and frustration. And he has the same potential for happiness as any other child. I've gone back to his note many times in the years since Henry was born. I remember my despair, and how moved I was by Michael's encouragement and understanding. Every time I read it, I hope that one day I'll write something so meaningful to another new parent caught in the throes of disappointment and fear.

Michael's message reminded me that I wasn't alone in the decisions I had made or the emotional turmoil I was feeling. My world, which had contracted to the lonely, impersonal space of a hospital bed, started to expand. Gradually, I found that being a mother to Henry didn't isolate me from other people, as I had feared. Instead, he became my bridge to an entirely different city than the one I had been living in for the past ten years. From there, he led me to a global network of people with Down syndrome, their families, and supporters. Although I've written on the subject of globalization, it was only as I learned to be Henry's parent that I came to understand, in a palpable way, what it means to belong to an interconnected world.

I started small. A few days after I got home from the hospital, I called our local chapter of the National Down Syndrome Society. The woman who answered the phone congratulated me on my new baby. This was a refreshing change from the doctors, who talked of our son only in terms of possible medical complications. She gave me the numbers of several families who had volunteered to serve as contacts for new parents. It turns out that when it comes to Down syndrome—as in so many other things—New Yorkers like to go it alone. Unlike nearly every other major city in the country, Manhattan has no organized parents' group. At first I was dismayed. I felt awkward about calling complete strangers. But when I finally got brave enough to pick up the phone, the women who answered were more than ready to talk. Over time I realized that there was a sizable community of sorts. Most of our contact takes place via e-mail, but occasionally we meet at informal gatherings, fund-raisers, or lectures and seminars.

It also turns out that New York City is a great place to be a young child with a disability. The Individuals with Disabilities Education Act (IDEA) guarantees Early Intervention services to all children diagnosed with developmental delays. Research has shown EI to be particularly effective for children with Down syndrome, whose prospects for development improve dramatically with early and consistent therapy. Because we live in New York State, which has one of the best EI programs in the country, Henry is entitled to treatment by a team of skilled therapists, who come to our apartment free of charge. I discovered that my city is home to an army of these therapists, almost all of them women, who spend their days traveling from home to home, lugging backpacks full of paperwork and equipment to treat their clients. For years I must have passed them on the sidewalk and shared seats with them on the bus, but Henry made them visible to me. Through these therapists, I've glimpsed a world beyond my own. They've taught me about innovative treatments coming from England and Israel. Our occupational therapist was learning to build equipment from cardboard, using techniques she hoped to export to Africa.

After imagining myself so completely alone, I was surprised to find that two boys with Down syndrome already lived on my street. And there were other families in my neighborhood. I wondered why I hadn't seen them before. Or maybe I had seen them, and then quickly looked away. Before I had Henry, children with Down syndrome represented a misfortune that had nothing to do with me. People with disabilities will tell you how common it is to feel like they're invisible. For the able-bodied, it's easy not to see the wheelchairs and the walkers and the crutches and the white canes, the people who limp, and those who are shorter or taller than average. Because of my academic interests, I've known this for a long time. But I still didn't *see*. Once your eyes are opened, a whole different world comes into view, one that most of us will join some day.

Our circle soon grew beyond our own small corner of Manhattan. Other families in our group come from all five boroughs, New Jersey, and Connecticut. The Internet plays a crucial role in connecting our scattered community. Like so many other dispersed communities in our globalized moment, technology holds us together. There was the mother who wrote to me in the middle of the night about how she sometimes hated the twin with Down syndrome, who was constantly spitting up and whose right ear folded over like a cabbage leaf. Another woman shared slideshows of her adult brother with those of us who wonder what our kids will be able to do when they grow up. Another used Facebook to send updates from the hospital where she and her son were living while he was being treated for leukemia. And there were all the useful discussions about where to buy glasses and orthotics, and who was willing to give swimming lessons to a kid with DS. I like to think that all of the Internet discussion boards, blogs, and Web sites might make it easier for prospective parents to choose to have a baby with Down syndrome. On my international listserv, families in Scotland and England connect with families in North Dakota and Pennsylvania. We are overjoyed for a mother in Los Angeles, whose daughter survived chemotherapy and kept her blood counts up for five months. Together we share indignation over the woman with Down syndrome who is asked to leave her college ceramics class because she needs extra assistance.

Of course it was all about the mothers. How could it be otherwise when women still do most of the child care in most parts of the world, whether or not they're fully employed outside the home? And when the child is disabled, women overwhelmingly shoulder the burden of any extra emotional and physical care. I wasn't surprised to find that mothers were the ones contributing to our e-mail list, going to fund-raisers, lectures, and information sessions, and organizing the social events. We're the ones who keep in touch and share resources. And we're the ones who enjoy the friendship and support that comes from belonging to our makeshift community.

When I finally met some of these women in person I was struck by how different we all are. There are mothers from Latin America, the Caribbean, Canada, England, South Africa, Holland, and India. We have widely varied economic resources, families, and career paths. We tell very different stories about how we came to have our babies. Many did so despite the disapproval of their doctors and other medical professionals responsible for their care. Some had religious beliefs that prohibited abortion. Others were like me, surprised by the diagnosis after their babies were born. Still others had received a "positive diagnosis" and just decided to continue their pregnancies, Down syndrome or no.

Some of us have a lot in common. One night I had dinner with a group that included two women who had known each other since their schooldays in Mexico City. By sheer coincidence, both had moved to the same neighborhood on the Upper East Side. And each had given birth to a child with Down syndrome. There's at least one other humanities professor in our group, and another woman with a PhD in Anthropology. There are also stay-at-home moms, teachers, lawyers, insurance agents, and a doula. Whatever our differences, we're united by the uncommon paths we've taken in a society where many regard the birth of a child with Down syndrome as a preventable mistake. We're drawn together by concern for the well-being of our children and our struggles with a world that is often resistant or hostile to giving them what they need to thrive.

Henry has introduced me to this other, more heterogeneous New York. But I'm constantly reminded that our world doesn't end at the borders of this city, or even the United States. Down syndrome has nothing to do with culture, region, or nation. It has no common demographic and is distributed across ethnic and socioeconomic groups. Although older women are more likely to give birth to babies with Down syndrome, it can happen to anyone.

Throughout much of history, inadequate medical care meant that many babies born with Down syndrome would not live past infancy. Sometimes they were allowed to die, even when care was available. And those who did were thought to be incapable of education. In the United States, parents were routinely advised to send them to institutions, where they languished without affection or stimulation.

Today there is ample evidence that, given proper resources and support, people with Down syndrome are capable of accomplishing a great deal. I keep a scrapbook that I know Henry will appreciate when he gets older. On the first page is a signed publicity photo of Chris Burke, the musician and actor who starred on the hit TV show *Life Goes On*. There are stories about the champion swimmer Karen Gaffney, who crossed the English Channel in the summer of 2001; the celebrated fiber artist Judith Scott; and the musician Sujeet Desai, who has performed around the world. In addition to these notable Americans, I also want him to know about people with Down syndrome in other parts of the world. I hope he'll feel inspired to learn that the tapestries of Canadian artist Jane Cameron hang in many different countries. And that Pascal Duquenne of Belgium, Pablo Pineda of Spain, and Paula Sage of Scotland have all won international film acting awards.

World Down Syndrome Day is March 21, a date that represents the tripling of the twenty-first chromosome that causes the condition. The twenty-five countries listed on the World Down Syndrome Day Web site span the globe, including Canada, Mexico, Japan, Singapore, Brazil, Saudi Arabia, Switzerland, Turkey, Kenya, and New Zealand. In 2010 they commemorated the occasion with lectures and symposia, fund-raising walks, performances, and parades. The fact that people with Down syndrome are celebrated in so many parts of the world represents a drastic change from past generations. It is a supreme irony that this greater acceptance of people with intellectual disabilities coincides with the development of ever more precise tests for detection and, presumably, elimination of "defective" fetuses. The impulse to search and destroy is pervasive throughout the Western world, and beyond.

Sometimes these contradictions come to a head. In 2008 I was dismayed by reports from Baghdad that two women with Down syndrome were strapped to remote-control bombs that detonated, killing at least 73 people and wounding 150. Although the women's diagnoses were later called into question, the disregard for the lives of people with Down syndrome that the story conveys is all too real. The Internet, which has been such a vital resource for sharing information and building community, also makes it easier than ever to spread hatred and prejudice. In early 2010, an Italian Facebook group called for children with Down syndrome to be used in target practice. It circulated a picture of a baby girl with the word "imbecile" written across her forehead. That same month, a Russian blogger wrote a widely circulated article titled "Finish it off so it doesn't suffer" that called for the euthanasia of disabled newborns.

This isn't to say that prejudice exists only abroad. Of course, there's plenty of homegrown bigotry to be found in the United States, beginning with the casual use of the offensive word "retard" in the halls of the White House, and in films like Ben Stiller's 2009 *Tropic Thunder.* The influential Princeton philosopher Peter Singer advocates killing infants with severe disabilities, and he has argued that individuals with some forms of cognitive disability are undeserving of moral consideration. In March 2010, the U.S. National Down Syndrome Congress had to dismantle a Web site featuring its "More Alike Than Different" campaign after vandals defaced several of the photos with cruel slurs. And when I was quoted at length for a story in *The Daily Beast,* one reader commented that I'm an irresponsible person. "We should make a social contract type deal on this topic," he wrote.

> Every single parent who knows they have a fetus with an extreme and expensive developmental disability such as Down syndrome is more than free to have the baby. The trade is that they should be responsible for the full cost of raising the child. That means no 70k from their school district for special classes, no aide, no special dispensations from the government, none of that. They made their bed, they lie in it. I bet you anything if we stopped socializing the cost of these choices, people would make different ones. It's easy to say "I want a child no matter how much it costs to treat him after he's born" when it's not your money.

His words remind me that there are still many people out there who think a child like Henry shouldn't be born, or that his life is expendable. I hope that someday, he'll be able to tell them how wrong they are. In the meantime, I consider it my job to represent his interests as best I can.

Just before Henry turned three, I saw *Yo También,* a Spanish film about Daniel, a man with Down syndrome, who falls in love with Laura, a nondisabled co-worker. It's a sensitive portrayal of the difficulties they encounter as his desire for adult intimacy conflicts with her desire for a nonromantic friendship. When their relationship reaches a crisis, Laura is forced to probe the sources of her resistance, confronting her own prejudice as well as her genuine love for Daniel. A parallel plot tells the story of a man and a woman with Down syndrome who attend the arts program where Daniel is employed. In their case, both desire the same thing—a mature relationship, sex included—but they have to contend with parents and caregivers who want to keep them in a perpetual state of childhood.

A number of things about the film reflect its foreignness. On the positive side, it portrays a society far more committed to making opportunities for meaningful work and artistic expression available to people with intellectual disabilities than my own. But I'm more skeptical about its suggestion that Down syndrome is something that can be mitigated or overcome through sheer force of will. When asked how he became so successful, Daniel explains that his mother pushed him to achieve where other parents would have accepted less from their children. The notion that Down syndrome can be conquered with a properly strict education is foreign to me. In an ideal world, I'd like to see each child receive the resources to develop to her fullest potential, but I'm also aware that children with Down syndrome—like all children—are greatly varied in terms of ability. It sends an unrealistic and potentially harmful message that each, with the right training, would be equally capable of the academic and professional success Daniel has achieved. A child who accomplishes less should not be seen as a sign of personal or parental failure.

Other subjects broached by *Yo También* are more universal. The themes of thwarted quest for intimacy and the challenges of adult sexuality among people with intellectual disabilities must resonate in any number of cultural contexts, since they reflect basic human needs and desires. Although my Henry is still a baby in many ways, I know that at some point in the future we'll need to contend with the questions about sex and relationships raised by the film. It left me thinking hard about the difficulty and importance of assuring that people with Down syndrome have opportunities to learn about mature, safe, and respectful forms of sexual expression. We'll need to walk a fine line between respecting Henry's privacy and independence, and making sure that he doesn't harm himself, or anyone else. These are intimate challenges that are global in nature. They must be shared by parents of children with intellectual disabilities in all cultures and regions of the world, although I have no doubt that we approach them with widely varied beliefs and attitudes.

The day after I saw *Yo También* I took Henry to the park. As soon as he got out of the stroller, he made a beeline for a group of bigger kids clustered around a fountain. They were loud and joyful and a little bit rough, engaging in exactly the kind of play he finds irresistible. I ran over, worried he might be knocked down. I noticed that somebody had already dumped water on his shirt and somebody else had grabbed his favorite pouring cup. Henry didn't seem to mind. In fact, he was having the time of his life, screaming with delight as he made his way directly into the center of the action. I felt the urge to hover, to make sure he got his turn at the water and that he didn't get hurt. But then I forced myself to step back, leaving him to look out for himself.

II

MEMORY, HISTORY, COMMUNITY

Personal

Narrative

in a

Transnational

Frame

5

FACING

INTIMACY ACROSS DIVISIONS

Mieke Bal

One of the most tenacious instances of universalism—the belief in the universality of something—is motherhood, doubtlessly the most intimate of relationships. The current state of the allegedly globalized world makes this universalism both urgently necessary and deeply problematic. This ambivalence is the topic of my contribution to this volume.

Thinking of motherhood as a universal usefully counters problematic relativizing. For example, relativizing the horror of losing a child by alleging that, in some severely underprivileged countries, losing a child to illness, hunger, or violence occurs so frequently that it is "normal" would be a painful condescendence and a scandalous acceptance of the unacceptable. In the globalized world, the opposite move is necessary. Now that we are bound to those underprivileged situations by knowledge and economic complicity, assuming the universality of motherhood—through a "strategic universalism"—is a political necessity. Only through that assumption does the scandal of the inequities that globalization both promotes and lays bare become apparent.

At the same time and even for the same reasons, the opposite move is just as indispensable. One of the most severe challenges to the idea, or hope, of any universality is the division produced all over the world between people whose everyday life and its intimacy are safely assured and those who lead an existence of "infra-humanity."[1] Among other consequences, this division also produces an unsettling tension when the two parts of our supposedly unified world collide in Western countries as a result of migration. Migration causes the coexistence in one social environment of people who can afford to live permanently in a place and those who cannot—those who are driven to displacement.[2]

This situation deeply impinges on motherhood. It interrupts that relationship and brutally destroys the relation of intimacy, since the proximity between mother and child is no longer a matter of choice. The combination of motherhood and migration, then, is a good place to reflect on the confrontation between globalization and intimacy against the backdrop of a nonoppositional binary of singularity and universality. The relationship between the singular and the general—to use a more abstract binary that encompasses both universalism-singularity and globalization-intimacy—also holds for my own analyses. It has consequences for the relationship between my video writing and my academic writing. Although one of my video installations is central to the argument here, this is not a traditional case study but an exploration, through one particular "case," of the dynamic complementarity between media. One goal is to make the mothers staged in this installation full participants in what can only be a multivoiced discussion. Another goal is to develop a methodology suitable for the object of study.[3]

Rather than generalizing on the basis of a singular case, I am constantly going back and forth between one special view and another. In terms of the logic of reasoning, this movement is neither deductive nor inductive but what the American philosopher Charles Sanders Peirce, with an idiosyncratic term, called "abductive." Jan van der Lubbe and Aart van Zoest define it as follows: "In general abduction is considered as that type of inference which leads to hypothetical explanations for observed facts. In this sense it is the opposite of deduction."[4] Abduction goes from consequence to possible cause. As Van der Lubbe and Van Zoest write, this type of logic is "diagnostic."[5] Deduction, in contrast, reasons from cause to consequence and is thus prognostic. Abduction is the way through which new ideas become possible. It makes creative leaps and has the singular as its starting point. It thrives on uncertainty and speculation, but its origin in observable fact remains primary. I consider abduction the most suitable form of reasoning in the face of globalization and the need to know what intimacy can mean for people at the other end of the economic division.[6]

This abductive approach to what appear to be case studies has been an increasingly stimulating guideline in my recent work. Over the past six years I have explored this tension through several video works on migration. Most of these are based on the performativity of intimacy with migrants; they are concerned with situations of displacement (*Lost in Space*) and show migrants struggling to achieve some level of integration (*A Thousand and One Days*; *A Clean*

Job) or suffering from the economic consequences of globalization (*Colony*). The tension between intimacy and the consequences of globalization is enacted most explicitly in a video installation made between 2006 and 2008. Through a discussion of this installation I seek to grasp intimacy on terms that allow for the strategic use of universalism ("motherhood") as well as for the foregrounding of differences ("migration").[7]

THE PROJECT

The installation titled *Nothing Is Missing* consists of a variable number of audio-visual units that play DVDs of about thirty minutes in which a mother talks about a child who has left in migration. Imagine a gallery looking like a generic living room, where visiting is like a social call (see figure 5.1). The image is a portrait, a bust only, of a woman speaking to someone else. Apart from a short introductory sequence that sets up the situation, the videos consist of unedited single shots. Sometimes, we hear the voice of the interlocutor; in other cases, we hear no one other than the woman speaking. Every once in a while, one of them falls silent, as if she were listening to the others. The installation itself enacts the tension between global and intimate, since the domestic ambience is created within a space that is public, although often not a space where such installations

5.1 *Nothing Is Missing* installation overview

are expected. I have installed it in museums and galleries, academic settings, and office spaces—in a corner office at the Department of Justice in The Hague, for example.[8]

This installation probes the contradiction between usages of universalism as escapist exclusion and as a strategy to enhance differences. My provisional answer to the contradiction between these two elements is to replace any thematic universalism with a performative one, and an essential universalism with a strategic dynamic variant that is constantly challenged by singularities. Between aesthetic and academic work, a certain activism through the promotion of reflection in sense-based experience is also present. The question that the video work raises, and that the present paper attempts to answer, is how it is possible to make intimate contact across the many divisions that separate people in different linguistic, economic, and familial situations, and why it matters to do so. The goal is not to reach a universal ground for communication but instead to establish the universal as the ground on which differences can performatively be brought into dialogue.

The women are from various countries from which people have migrated since the onset of modern-day globalization. Still living in their home countries, they all saw a child leave for Western Europe or the United States. My project is not an attempt to understand migration as such, nor to defend its necessity, which I take for granted. Rather, if we are to understand the possibility of a universal such as motherhood through insight into intimate relationships against the backdrop of a globalized world, we must first of all realize the enormity of the consequences involved and the changes in the souls of individuals taking this drastic step. We must wonder, that is, why people decide they must leave behind their affective ties, relatives, friends, and habits—in short, everything that constitutes their intimate everyday lives. These motivations, which are too complex to allow any generalizations, tend to include economic necessity but are rarely limited to that overarching issue. While my purpose is not to fully understand those complex motivations, I bring them to the fore here, considering that they are relevant in being among the ambivalences toward the migration of their child to which the mothers testify. My primary goal is to explore the possibility of an "aesthetic understanding" that, by means of its own intimacy across the gaps of globalization, can engage the political.

These terms refer to a simple understanding of the two domains. For aesthetics I return to the eighteenth-century philosopher Alexander Gottlieb Baumgarten, who developed the notion of binding through the senses, and, incidentally, also considered aesthetics a useful approach to the political.[9] Since this conception presumes neither beauty nor a separate artistic sphere, it seems a useful starting point to develop the idea of an aesthetic understanding that straddles the distinction between academic and artistic exploration. Moreover, the proximity presupposed by the sense-based experience also establishes intimacy between the subject and the object of the aesthetic moment. Hence, this approach furthers my attempt to develop a methodology that approximates the "object."

For the political I rely on the distinctions between politics and the political currently advanced by, among others, Jacques Rancière[10] and Chantal Mouffe. In a clear and concise book about this distinction, the latter defines the two terms as follows: "By 'the political' I mean the dimension of antagonism which I take to be constitutive of human societies, while by 'politics' I mean the set of practices and institutions through which an order is created, organizing human coexistence in the context of conflictuality provided by the political."[11]

In this distinction, politics is the organization that settles conflict; the political is where conflict "happens." Thanks to the political, social life is possible. Politics, however, constantly attempts to dampen the political. Rancière uses different terms for the same distinction. In his work, Mouffe's "politics" corresponds to "the police," and her "political" is identical to his "politics." Since I find Mouffe's terms clearer, I will proceed to use those.

According to Mouffe's view, everyday life, including the intimacy that inhabits it, pertains to the political. It is there that intimacy must be grasped. In my view, a first step to contemplating these questions is a triple act of *facing*. Facing sums up the aesthetic and political principle of my video work *Nothing Is Missing*, which is an attempt to reflect on severance and its consequences. Through this installation, I attempt to shift two common universal definitions of humanity: the notion of an individual autonomy of a vulgarized Cartesian *cogito* and that of a subjecting passivity derived from the principle of Bishop Berkeley's "to be is to be perceived." The former slogan has done damage in ruling out the participation of the body and the emotions in rational thought. The latter, recognizable in the Lacanian as well as in certain Bakhtinian traditions, has sometimes overextended a sense of passivity and coerciveness into a denial of political agency and, hence, responsibility. Reflecting on facing helps me to rethink these notions. I try to shift these views in favor of an intercultural, "relational" aesthetic based on a performance of contact. In order to elaborate such an alternative I have focused this installation on the bond between speech and face as the site of the performance of a universal. Here, I use speech not just in terms of "giving voice," but also in terms of listening and answering, all in their multiple meanings; furthermore, I would like to turn the face, the classical "window of the soul," into an "inter-face."

FACING PHILOSOPHY

Facing constitutes three acts at once. Literally, facing is the act of looking someone else in the face. It is also coming to terms with something that is difficult to live down by looking it in the face rather than denying or repressing it. Third, it is making contact, placing the emphasis on the second person, and acknowledging the necessity of that contact simply in order to sustain life. Instead of "to be is to be perceived" and "I think, therefore I am," facing proposes, "I face (you); hence, we are." For this reason, facing is my proposal for a performance of contact across divisions that avoids the traps of universalist exclusion and relativist condescendence.

For this purpose, I first make the move from the two universalist views of humanity—Descartes's and Berkeley's—to a merger that replaces both; from *esse est percibi* to *cogitote ergo sumus*. There is no clearer—almost programmatic—demonstration of Berkeley's view than Samuel Beckett's only film, called *Film*. As Anthony Uhlmann has pointed out, Berkeley's formula, as elaborated to exhaustion by Beckett, is agony-inducing.[12] As it happens, this identity without agency already shows linguistically in the mere fact that the formula defines being in nonpersonal forms. As a result, Uhlmann argues, Beckett's film explores the agonizing feelings that result from a consciousness of being through being perceived.

Film explores the relationship of disharmony between the three types of images Gilles Deleuze distinguishes in *Cinema 1*. The perception-image is the result of the viewer's selection from the visible world of those images that might be useful for her. The action-image presents possibilities to act upon what is seen. In between, the affection-image compels the viewer—who is affected by the perception—to consider action. Stuck in (negative) affect when he is the object of someone else's perception, the protagonist of *Film*, played by the aging and decidedly *not* comical Buster Keaton, flees from the notion of perceivedness in the film's "action-images." The sets of eyes that watch him and that he eliminates show us the violence of the "perception-image," whereas the ending, the close-up of the "affection-image," translates affect into pure horror. This story can offer a useful counterpoint for the installation *Nothing Is Missing*. There, these three types of images culminate in the mitigated close-up of the face that

shuttles between perception-image and affection-image without the leap to action. Here, neither horror as a form of revolt, nor passive perceivedness as a
handing over of human agency, but a rigorously affirmed second-personhood
is the reply to this pessimistic view. The perceivedness that the predominance
of the close-up foregrounds does not lead to either rejection or agony, but instead
to an empowering performativity. This, then, is my reply to Berkeley's pessimistic view of vision as violence.

Now, Descartes. The notion that Descartes is the bad guy of Enlightenment
rationalism seems to reduce him in the same way as he was seen to be reducing
human existence. According to Jean-Joseph Goux, the stake of the cogito is not
primarily the link between thinking and being, nor even the exclusive emphasis
on reason and the excision of the body, but the tautological grammatical use of
the first person: *I* think, [therefore] *I* am.[13] The point is the possibility to describe human existence outside of the need to use the second person.

The popularity of his formula has done more harm than good to Western
thought, especially in its exclusions, its excising of not only emotions but also
the dependency of human life on others from human existence. I call it an *autis-
tic* version of humanity, and deny it the universality it has come to claim. Yet the
dependency on others is so obvious and absolute that it may well have been its
very inevitability that informed the desire to erase it in the first place. From the
baby's mother to social caretakers to linguistic second persons, this dependency
has been articulated clearly in psychoanalysis, sociology, and linguistics, respec-
tively: so much so, in fact, that being a second person seems more "natural" a
definition of being human than anything else. Second-personhood, I contend,
may well be the only and most important universal of human existence, while
its repression underlies other universalist definitions.[14]

This means that we cannot exist without others—in the eye of the other as in
the eye of the storm (Berkeley, Beckett), as much as in sustenance of others (the
ethical imperative to which Descartes refuses to owe his existence). That is
where I would start any attempt to confront universality as the ground where
globalization meets—allows, enables, or precludes—intimacy. I do this not to
pursue the beating of the Cartesian dead horse, but, on the contrary, to keep in
mind the productivity of returning with "critical intimacy"[15] to moments of the
past, such as the dawn of rationalism in the seventeenth century. In this I am
joining a growing group of scholars exploring the history of thought and developing alternative ways of thinking humanity, many influenced by Deleuze, his
Spinoza, his Leibniz, and his Bergson—to *name* the names that underlie my
thoughts here.[16]

An increasing number of scholars are studying the relevance of Descartes's contemporary Baruch Spinoza for an alternative stream of thought between early and late rationalism. The line Spinoza—Bergson—Deleuze has led to extremely important and productive revisionings of the image, perception, and feeling. Some of these new ideas lie at the heart of the "migratory aesthetics" of my installation—an aesthetics of geographical mobility beyond the nation-state and its linguistic uniformity. Philosophers Moira Gatens and Genevieve Lloyd wrote a useful book that unpacks those ideas in Spinoza's writing that can be employed for contemporary social thought.[17] Gatens and Lloyd's book does three things at once that are relevant for my project, furthering the activity of "migratory aesthetics" and deploying the performative face in that context.

First, they develop an intercultural relational ethics. To this end, they invoke the relevance of Spinoza's work for a reasoned position in relation to aboriginal Australians' claim to the land that was taken from them by European settlers. These claimants are not migrants since they stayed put while their land was taken away from underneath them, but their claims are based on a culturally specific conception of subjecthood and ownership that makes an excellent case for the collective and historical responsibility the authors put forward with the help of Spinoza. This responsibility is key to any possible universality. It is a relation to the past that we have to face today.

That this intercultural ethics should be based on a seventeenth-century writer who never met such claimants—although he was definitely a migratory subject—makes, second, a case for a historiography that I have termed *preposterous*.[18] This conception of history is focused on the relevance of present issues for a re-visioning of the past. In alignment with intercultural relationality, we could call it *inter-temporal*. Third, the authors make their case on the basis of the integration, an actual merging, of Spinoza's ontological, ethical, and political writings—three philosophical disciplines traditionally considered separately. This, of course, exemplifies interdisciplinarity. In order to transform it from a fashionable buzzword into an intellectually responsible and specific notion, interdisciplinarity could be modeled on inter-facing in the sense I am developing here: as a universalist practice.

Against this background—my search for an alternative to masochistic passivity and autism as a ground for the possibility of a performative universal—the face, with all the potential this concept-image possesses, seemed an excellent place to start. But to deploy the face for this purpose requires one more negative act: the elimination of an oppressive sentimentalist humanism that has appropriated the face for universalist claims in a threefold way—as the window of the

soul, as the key to identity translated into individuality, and as the site of policing. With this move I also seek to suspend any tendency to sentimentalizing
interpretations of *Nothing Is Missing*.

IDEOLOGIES OF THE FACE

The abuses of the face that individualism underpins are, in turn, articulated by
means of a form of thought that confuses origin with articulation, and runs on
a historicism as simplistically linear as it is obsessive. Common origin is a primary ideology of universalism. This involves motherhood: all human beings are
born from a mother (even if this universal is no longer true). Creation stories
from around the world tend to worry about the beginning of humanity in terms
of the nonhumanity that precedes it. Psychoanalysis primarily projects on the
maternal face the beginning of the child's aesthetic relationality. Both discourses of psychoanalysis and, as I will demonstrate shortly, aesthetics show
their hand in these searches for beginnings. Both searches for origins are predicated on individualism, anchored as they are in the mythical structure of evolution as ongoing separation, splitting, and specification.

Here I take issue with an individualistic conception of beginnings through
an alternative view that I will draw from literary theory. A few years after his
path-breaking book *Orientalism*, the late Palestinian intellectual Edward Said
wrote a book on the novels of the Western canon, entitled *Beginnings: Intention
and Method*. In this book he demonstrated that the opening of a literary work
programs the entire text that follows, from its content and its style to its poignancy and aesthetic. It is the thesis of *Nothing Is Missing* that this is true for
cultural-political reality as well. Origin is a forward-projecting illusion. Therefore, in this installation I wished to explore a different sense of beginning—not
in motherhood, but in migration. The primary question is why people decide
to leave behind their lives as they know them and project their lives forward
into the unknown. With this focus, I aim to invert the latent evolutionism in the
search for beginnings, and, in the same sweep, the focus on children and babies
inherent to that strange contradiction of individualistic-universalist theories of
the subject.

Today, with authorities displaying high anxiety over the invisibility of the
Islamic veiled face, we cannot overestimate the importance of the ideology of
the face for the construction of contemporary socio-political divides. To briefly
show the workings of this ideology I look to an art-historical publication that

earned its stripes in its own field: a study on the portrait, the artistic genre par excellence, in which individualism is the *conditio sine qua non* of its very existence.

Confusing, like so many others do, origin with articulation in his study of the portrait—the genre of the face—art historian Richard Brilliant explains the genre with reference to babies: "The dynamic nature of portraits and the 'occasionality' that anchors their imagery in life seem ultimately to depend on the primary experience of the infant in arms. The child, gazing up at its mother, imprints her vitally important image so firmly on its mind that soon enough she can be recognized almost instantaneously and without conscious thought."[19]

Like psychoanalysis, art history here grounds one of its primary genres in a fantasmatic projection of what babies see, do, and desire. Both disciplines can and must be challenged for their universalism.[20]

The shift operates through the self-evident importance attributed to documentary realism, a second unquestioned value in Western humanist culture that has been elevated to a universal status and that has also been inscribed in the face. Identity pictures as a form of policing demonstrate the bond between these two sides of the ideology of the face. The point of the portrait is the belief in the real existence of the person depicted, the "vital relationship between the portrait and its object of representation."[21] The portraits that compose *Nothing Is Missing* challenge these joint assumptions of individualism and realism and their claim to generalized validity.

The women in this work are, of course, "real," as real as you and me, and individual—as different from you and me as the world's divides have programmed. At first sight, they have also been documented as such. At the same time, however, the installation enables them to speak "together" from within a cultural-political position that makes them absolutely distinct and absolutely connected at once. This is the meaning of the silences that suggests they are listening to one another, even if they never actually met.

As for the documentary nature of their images, again, this is both obvious and obviously false, since the situation of speech is framed as both hyperpersonal and utterly staged. I filmed the migrants' mothers talking about their motivation to support or try to withhold their children who wished to leave and about their own grief to see them go (figure 5.2). The mothers talk about this crucial moment in their past to a person close to them, often someone whose absence in their life was caused by the child's departure—a grandchild, a daughter-in-law, or the children themselves. This is a first take on the universal performance of contact I want to propose, against the more exclusionist universalities.

she got married
and now lives in america

5.2 Samira

In this performance, I contend, intimacy plays itself out against the odds of globalization-informed separation.

The act and mode of filming itself is implicated in this theoretical move. It is, in one sense, perfectly and perhaps excessively documentary. I staged the women, asked their interlocutors to take place behind the camera, set the shot, turned the camera on, and left the scene. This method is hyperbolically documentary. To underline this aspect I refrained from editing the shots. I will return to the resulting slow, unsmooth, and personal talk that results.

Aesthetically, the women are filmed in consistent close-up, as portraits—the other side of the face of Brilliant's babies. The relentlessly permanent image of their faces is meant to force viewers to look these women in the face and listen to what they have to say, in a language that is foreign, using expressions that seem strange, but in a discourse to which we can relate affectively. This is a second form of the performance of contact. Another assumption of Brilliant's argument concerns the *nature of identity*. In his view, identity is based on the baby and is enabled by seeing the mother's face; in this way, the ontogenetic perspective is constantly mapped on the phylogenetic one, in which development is the matrix and old equals primitive. This baby-basis is challenged most explicitly by the simple fact that the figures speaking here are the mothers, the other side of that face gazing up at them; they now become the holders of the inter-face. The face as inter-face is an occasion for an exchange that, affect-based as it may be, is fundamental in opening up the discourse of the face to the world.

Crucially, for Brilliant, identity emerges not only out of appearance and naming but also out of distinction. Moreover, the recognition of appearance triggers interaction and expression. Typically for the cogito tradition, the two are practically the same:

> Visual *communication* between mother and child is effected face-to-face and, when those faces are smiling, everybody is happy, or appears to be. For most of us, the human face is not only the most important key to *identification* based on appearance, it is also the primary field of expressive action.[22]

The assumed link between these two sentences equates communication with identification and expression. This equation is grounded in the double sense of identification—*as* and *with*—that underlies the universalist paradox and to which my installation attempts to consider an alternative. I call that alternative "inter-facing."

The socio-cultural version of this political ambiguity is most clearly noticeable in the dilemma of "speaking for" and the patronizing it implies, versus "speaking with" as face-to-face interaction. The self-sufficient rationalism of the cogito tradition is thus in collusion not only with a philosophical denial of second-personhood but also with a subsequent denial of what the face, rather than expression, can *do*. In order to move from an expressionism to a performativity of the face that, I contend, writes a program for a new, tenable universality, I deduce three uses of the preposition "inter-" from Gatens and Lloyd's take on Spinozism that can be mobilized in a helpful way. But in order to prevent an overhasty, overoptimistic mystification, each "inter-" works across a constitutive gap.

INTERCULTURAL ETHICS: RELATIONALITY ACROSS GAPS

Inter-*cultural* relationality, in its inscribed mobility of subjectivity, posits the face as an interlocutor whose discourse is not predictably similar to that of the viewer. These women speak to "us," across a gap, as they speak to their own relatives, again across a gap. The first gap is that of culture, if we continue to view cultures as entities instead of processes. In such a conception, intercultural contact is possible on the basis of the acknowledgment of the gap that separates and distinguishes them. The sometimes overextended emphasis on difference in postcolonial thought is a symptom of that gap. The second gap is caused by "the cultural" conceived as moments and processes of tension, conflict, and negotiation.[23] To highlight this dynamic, including the gap, I have chosen to invite the mothers to choose intimates as the interlocutors. The people to whom the women tell their stories are close to them, yet distanced by the gap that was caused by the migration of the loved one. Massaouda's daughter-in-law, for example, who was not chosen by her, is reaching out to the mother across an unbridgeable gap produced by history. In Elena's case, even the son himself struggles to overcome the gap that sits between him and his mother, with whom he talks during the short summer period when he visits her (figure 5.3).

There are yet other gaps in play. As I have suggested, the two simultaneous situations of speech—between the mothers and their relatives and between the

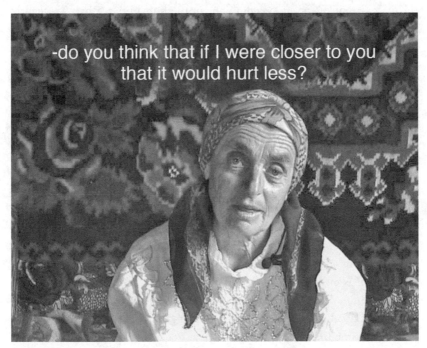

5.3 Elena

mothers and the viewer—doubly mark second-personhood, but across a gap. The strong sense of intimacy emanating from the direct address of the mother to her closely familiar interlocutor at first excludes the viewer. Only once one makes the effort and gives the time to enter the interaction can the viewer earn a sense of participation. When this happens—and, due to the recognizability of the discourse, it does—the experience is exhilarating and, I contend, unique in public events such as art exhibitions.[24]

The third gap opens in the making, due to the theoretical and artistic alternative to artistic authority of a "willful abandon of mastery," which underlies the filming in my own absence. There is *necessarily* a gap between intention and artwork. The gaps as entrances into sensations that are "borrowed" in a sense, grounded in someone else's body, open the door to the inter-face. Gaps, in other words, are the key to a universality that rejects a romantic utopianism in favor of a difficult, hard-won, but indispensable inter-facing. Gaps, not links, are also the key to intermediality. As my installation attempts to suggest, the two go hand in hand.

There is another discourse to be addressed here, in the wake of the humanism à la Brilliant and the self-enclosure of vulgarized Cartesianism. I am referring to the discourse of *intention*, predominant in the humanities. I have frequently argued against the relevance and tenability of that discourse, most extensively in *Travelling Concepts in the Humanities*. The theoretical and artistic alternative of a "willful abandon of mastery" I have discussed there underlies the filming in my own absence that I practiced in the videos for this installation.

Uhlmann points this out through Beckett, and the latter uses that same noun, *gap*: there is *necessarily*, not coincidentally, a gap between intention and artwork. Beckett wrote this in a rare joyful passage, where he describes the sense of accomplishment he felt precisely because of his failure to do what he had intended: "I felt it really was something. Not quite in the way intended, but as a sheer beauty, power and strangeness of image.... In other words ... from having been troubled by a certain failure to communicate fully by purely visual means the basic intention, I now begin to feel that this is important and that the images obtained probably gain in force what they lose as ideograms."[25]

On a profound level of interdiscursivity, this statement engages the question of intention as medium-specific, binding it to images and, hence, bringing the discussion of intention within the domain of art history, where the relevance of intention is usually not discussed but taken for granted.

For someone like Beckett, who was a writer before all, this serendipitous experience was crucial. Uhlmann concludes his essay with the following summary of what, in the wake of the affiliation he establishes between Berkeley, Bergson, and Beckett, the image does to intention. I quote this formulation because it succinctly sums up why the image is productively incompatible with intentionalism—an incompatibility that, I argue, is most useful for a migratory aesthetics of the face. He writes: "What *Film* in part offers is the exploration of a medium that draws its power—the power to produce sensations—through gaps. Yet, images provide sparks that leap from one side to the next, like messages across synapses, thereby allowing the formation of a unity among difference: intuition and sensation, intuition and the idea, intention and reception, philosophy and literature."[26]

Significantly and paradoxically, Uhlmann uses the discourse of medium-specificity here to make a point about the merging of domains and the discourse of embodiment—sensations—to posit gaps. The gaps as entrance into sensations are grounded in someone else's body, opening the door to the inter-face. Gaps, in other words, are the key to a migratory aesthetics that binds globalization to a transformed intimacy.

PREPOSTEROUS TIME

This concept of the gap lays the ground, in turn, for the second partner in the exploration of "inter-"—namely, inter-temporal thinking, which comes with the preposterous foregrounding of the present as starting point. These women carry the history of the severance from their beloved child. They state their acceptance of that separation as a fact of the present. Moreover, the concept of video installation positions the co-presence of the mothers with the viewer visiting the installation. Here lies one function of the acoustic gaps, the silences in the films. When they do not speak, it seems as if it is the viewer's turn to speak back to the mothers, who are now just looking the viewer in the face.

The inter-temporality also plays out in the belatedness of the viewer's engagement. To understand the need for this engagement in its inevitable belatedness, two distinct steps need to be taken. The first makes the move from individual to social, the second from past to present. At the same time, the social nature of intersubjectivity holds the performative promise of the improvement of the social fabric that the imaginary enactment of identification will help to build.

The images themselves fulfill a function in this inter-temporality. They do this through the exclusive deployment of the close-up as affection-image. Here,

Spinoza's writing on affect becomes relevant. As Gatens and Lloyd recall, "The complex interactions of imagination and affect [yield] this common space of intersubjectivity, and the processes of imitation and identification between minds which make the fabric of social life."[27]

Aesthetic work may be eminently suitable to double-bind the women to a social world whose fabric allows their experience to be voiced. Instead of being caught in a double-bind that forces them into silence, they can be relieved of carrying their burden too solitarily. This is where the affection-image, which Deleuze theorized as emblematically situated in the close-up, comes in with its typical temporality. Close-ups subvert linear time. They endure and thus inscribe the present into the image. Between narrative images and close-ups, then, a particular kind of intermediality emerges: one that stages a struggle between fast narrative and stillness. Here, the type of inter-temporality that is at stake takes the present of viewing as its starting point.

Paola Marrati points to the crucial function of the affection-image as the closest to both the materiality of the image and that of subjectivity. She writes: "Between a perception that is in certain ways troubling, and an action still hesitant, affection emerges."[28] The affection-image binds a perception that has already taken place but leaves a trace to the future of possible action. This is why the affection-image remains closest to the present while providing it with the temporal density needed to make the inter-face possible.

Gatens and Lloyd recall that Spinoza's conception of affect is explicit in its inter-temporality. They write: "The awareness of actual bodily modification—the awareness of things as present—is fundamental to the affects; and this is what makes the definition of affect overlap with that of imagination. All this gives special priority to the present."[29]

The resulting images are far from the documentary realism so dear to Western culture. They possess a temporal density that is inhabited by the past and the future, while affect (and especially the affect produced by the close-up) remains an event in the present—an event of, to use a typical Spinozan-Deleuzian phrase, *becoming*. This is not an event in the punctual sense, but a slice of process during which external events slow down or even remain out of sight. Becoming concerns the presence of the past. If we take this presence to the realm of the social, we can no longer deny responsibility for the injustices of the past, even if we cannot be blamed for it. Without that responsibility, the use of the vexed pronoun "we"—"the full deceptiveness of the false cultural 'we'"[30]—itself becomes disingenuous, even unethical.

Gatens and Lloyd's "Spinozistic responsibility," then, is derived from the philosopher's concept of self as social, and consists of projecting presently felt responsibilities "back into a past which itself becomes determinate only from the perspective of what lies in the future of that past—in our present." Taking seriously the "temporal dimensions of human consciousness" includes endorsing the "multiple forming and reforming of identities over time and within the deliverances of memory and imagination at any one time."[31] This preposterous responsibility based on memory and imagination makes selfhood not only stable but also unstable.[32] This instability is a form of empowerment, of agency within a collectivity-based individual consciousness. In Deleuze's work, this becomes the key concept of becoming.

FACING RESTRAINT

Becoming also defines our activities as scholars in the humanities. Hence, finally, inter-*disciplinary* thought is needed. This allows us to make the connection, in the present and across the cultural divide, between a number of discourses and activities routinely either treated separately or unwarrantedly merged. It may be somewhat surprising that, in the course of this project, I became more cautious with the self-evident value of any form of interdisciplinarity. I have been a fervent proponent of interdisciplinarity for a long time. From the women in *Nothing Is Missing* I have learned what I had only intuited earlier: sometimes, invoking a disciplinary framework can do more damage than good to the insights we try to develop through it.

There are many issues here, but I will focus on a single one. The most obvious case seems also the most problematic one: the place of psychoanalysis, the darling approach of some and a changeling for others. I was faced with the need to hold back in this respect. Obviously, I do not dismiss the theory. But, lest I universalize Western conceptions, in some cases it was necessary to give full weight to the mothers' enacted desire to refrain from self-expression.

First, the situation of filming, in the intimacy-with-gaps and in the absence of the filmmaker, could easily become a trap to solicit more self-expression than the women would want to endorse. The intimacy of the speech situation has a globalized world of viewing as its backdrop, after all. But it is at moments of restraint, when they seem most reluctant to express themselves (in the Western sense of that phrase), that the *performativity* of their self-presentation is most acutely able to pierce through the conventional surface. These are the moments of the performative inter-face. I will describe one instance where the "performance of reticence," so to speak, in fact yielded the most beautiful insight into the way intimacy and globalization intersect.

The woman I filmed first, Massaouda, offers a striking instance of a culturally specific reluctance that cautions us against psychologizing or psychoanalyzing her. Not coincidentally, this occurs at the most strongly performative moment of the video. This is the situation: as I have been able to see first-hand, Massaouda and her newly acquired daughter-in-law, Ilhem Ben-Ali Mehdi, get along famously. But in their relationship remains the stubborn gap that immigration policies have dug. When Ilhem married Massaouda's youngest son, the mother was not allowed to attend the wedding: the authorities had denied her a visa. Not only had Massaouda not been granted the opportunity to witness who Ilhem was, but, even more obviously, she had not been able to fulfill her motherly role as her culture prescribes it, which is to help her son choose his bride. At some point, Ilhem ends up asking with some insistence what Massaouda had thought of her when she first saw her, after the fact and, hence, in a situation of powerlessness.

First, Massaouda does not answer, which makes Ilhem anxious enough to insist, and to ask: did you find me ugly, plain? The older woman looks away at this point. The young woman insists. We will never know what Massaouda "really" felt, but the power that the filming bestows on her, as if in compensation for her earlier disempowerment, is to either withhold or give her approval. She does the latter, but only after some teasing. When I saw the tape and understood the speech, I was convinced Ilhem would normally never have been allowed to ask this question and thus vent her anxiety—an intuition she later confirmed. As for the mother, she was given and then exercised the power she had been denied, and she used it to first mark the gap, then to help her somewhat insecure daughter-in-law.

"We"—global, mostly Western viewers of adult age—can easily relate to this moment. Such insecurity, for example, can easily be construed as universal. This interaction between Massaouda and Ilhem is thoroughly social, performative, but also bound to the medium of video—to the making of the film. Yet it does not allow, say, a universalizing psychoanalytic interpretation. Neither did I as maker have any influence on this occurrence—it was not my "intention." Nor can we construe it as a realistic, documentary moment in the sense of Brilliant's Gadamerian analysis of portraiture, where an "occasion" was recorded—it would never have happened outside of the situation of video making. Thus, it contradicts and suspends the universalizing myths of realism and documentary "truth." There would never have been an external reality the film could have documented. It is a moment, in other words, that was staged, yet real, thus challenging that distinction. Nor can we pinpoint a psyche offering symptoms for interpretation. For this to happen there was, instead, a need for a culturally specific relationship between two women related by marriage and separated by the gaps of migration, and for a relationship to the medium that allowed the women to overstep cultural boundaries.

Thus, reflecting on what I have learned from this experimental filmmaking, I felt compelled to extend my willful abandonment of mastery from the filming to the critical discourse I am offering here. An installation of voices, intermingling and alone—of women facing other women none of them had ever seen: I did this, but I could not master how I did it. The art making, in other words, is not an instance, an example to illustrate an academic point, nor an elevated form of cultural expression. Instead of these two things, equally problematic for a productive confrontation of universalities, I propose the universal validity of the performance in its nonuniversal singularity, including the moment of slight tension between Massaouda and Ilhem. The performative moment is the product of an act of filmmaking that required the absence of the filmmaker.

Moreover, it also required the surrender of the two women to the apparatus standing between them. This surrender entailed a cultural transgression—to insistently ask a question that in the culture of origin would be unspeakable. This, more than her linguistic pronunciation of Arabic as a second language, is Ilhem's "accent," in the sense in which Hamid Naficy famously uses that term.[33] This "accent" emblematizes the productive, innovative, and enriching potential of intercultural life. In this case, it could occur thanks to the absence of the filmmaker—but also of the two husbands—and the situation of displacement for both women. This interaction—between the people performing and the critic reflecting on how to understand what they did—would be stifled if an overly familiar psychoanalytic apparatus were let loose on this event.

This is as useful a lesson for a scholar interested in interdisciplinarity as any. It takes us out of the somewhat despairing "anything goes" posture that the flag of interdisciplinarity seems to cover too often (and which the indifferent use of the term "multidisciplinarity" betrays). The insight is the result of the shift from an essentialist concept of a static culture to a performative, confrontational concept of what could be called "the cultural." In this adoption of Fabian's concept of culture as a process of contestation and in analogy to Mouffe's distinction between politics and the political, I see a possibility to articulate an intimate cultural dynamic in the globalized world: the intercultural, indeed.

FACING SPEECH

Massaouda's and Ilhem's performances of intercultural contact were done on the basis of a close collaboration of the face and the word. Indeed, the spoken word is central to a performance of contact across divisions as well as to the installation. The word is deployed in an attempt to turn a condescending act of "giving voice" into an affirmation of our need to be given that voice. More directly than film, video binds the image we see to the sound we hear. That sound is, in this case, primarily and almost exclusively the human voice and the spoken words it utters. Speech, then, becomes the occasion for a positive deployment of interdisciplinarity, one that operates through intermediality.

Firstly, the centrality of the spoken word impinges on the visual form, the close-up. Film studies have been keen on including sound in their analysis,[34] but the visual appearance of words in subtitles seems to solicit nothing but indifference, both in the film industry, where the ugliest outlining of words pollutes the most beautiful images, and in the work of scholars who tend to ignore that aspect. In *Nothing Is Missing* I have attempted to experiment with the visualization of speech in order to make the most of the convergence of words and images. The surtitles, for example, make it easier to read the words and watch the faces at the same time.

In order to further privilege the voice of the mothers, the films consist of single unedited shots of their faces as they speak and listen. The personal situation presupposes sincerity. At the same time, they are keenly aware of the public nature of the speech they are producing in front of the camera. The nature of this performance is closer to theatricality, in the critical sense, than to traditional filmmaking. As theater, the situation is closer to minimally rehearsed, improvised, and inquiring forms of theater than to perfectly mastered public forms.[35]

Secondly, the translations presented as surtitles also embody the close bond between the linguistic and the visual aspects of the images—the bond between face and speech. As I mentioned earlier, the viewer is confronted with different languages, foreign to most, audible in their foreignness and visible in an emphatically visualized translation. Placed above their faces, the language is both made important and presented as somewhat of a burden. English as the universal entrance port is exploited as well as de-naturalized, both by this visual foregrounding and by the translations themselves.

Translations are as literal as possible, bringing out the poetry in the original languages without sacrificing to clarity. None of the translators are native speakers of English. Their assignment was to help me stay as close as possible to the phrasings the women used. This method results in this "accented" English that maintains the bi-cultural status of the communication.

Finally, the most acute intermediality occurs in the faces, which visibly produce the sound of the voices through their movement, thus yielding the movement of the image by means of sound. For this, with the language we do not understand, and the need to translate, all in one, the face is the actor. It is really difficult to separate sound from vision, since the mouths articulate with the rhythm of the sounds. This is not simply a case of the "moving image" of cinema. Instead, the moving quality becomes a poetic, a self-reflective statement about the medium that re-integrates what the predominance of English as universal language has shattered. This stands in contrast to the particular home-boundedness resulting from a lack of education, in turn aggravated by misogyny and colonialism. In this way, the face and its acts become the emblematic instance of video's power to transgress boundaries of a variety of kinds.

CONCLUSION

In *Nothing Is Missing* I do address actual migration, but not as the thematic heart of the work. That heart, rather, is the encounter with the faces as negotiated universality, where globalization meets and inflects intimacy—and vice versa. The focus on the face embodies the act of facing in its three meanings, all three staged here as acts of mutuality facilitating contact. First, the emphasis on activity reflects back on the face itself. No longer the site of representation and expression, the face has become an agent of action: what this installation demonstrates is what faces can do rather than how to do things with faces.

How, then, can the face be a universal, without presuming that facial expression is cross-culturally present or stable? The face faces, looking us in the face, which makes the viewer the interlocutor. It faces something that is hard to live down—here, the severance of the primary bond that humanism construes as defining for humanity: that between mother and child. In these videos of acting faces, that event is qualified as larger than the individual. All women speak in understated tones of the causes of the child's departure, and they do so in terms for which Western cultures can assume some measure of historical responsibility, if only "we" reason with Spinoza.

The severances, all having different causes in the past and being experienced differently in the present, are lived as what for me is the ultimate tragedy: that all of the mothers say they are *happy* about the sore fact that their child left. These backgrounds are understated because they can neither be eliminated from the present nor be allowed to overrule the existence of the mothers in an everyday that is also rich and sometimes happy. Hence, the discourse intimated in the installation's title—the one on which Massaouda ends her eventual and hard-won openness about what matters most to her as a mother: that her son finds bread to eat. Facing these present pasts, this kind of recognizable and perhaps, dare I say the word, universal motherhood that results nevertheless fulfills the becoming of who we are in the present: facing these pasts together so that "we" can "be" is part of our own potential of becoming.[36]

But how can we do that? Making contact, the third and most important act implied in facing, facilitates that becoming—becoming world citizens, building our existence on mobility without having to move. This making of contact is suggested as an effect of the insistent facing in *Nothing Is Missing*. What faces can do is stage encounters. This is the point of the mothers' faces in *Nothing Is Missing*—their empowerment. In the installation, the face is constantly present, in close-up but not as close as possible. As a visual form, the close-up itself *is* the face: "There is no close-up of the face. The close-up is the face, but the face precisely in so far as it has destroyed its triple function [individuation, socialization, communication] . . . the close-up turns the face into a phantom . . . the face is the vampire.[37]

If the close-up is the face, the face is also the close-up. Hence the slight distance nevertheless built into the image to avoid locking the viewer up and denying the women any space at all: To avoid facile conflation and appeals to sentimentality. To give the face a frame within which it can exercise its mobility and agency. To make the images also look a bit like the busts of Roman emperors and other dignitaries. That slight distance, then, provides the space for a certain kind of freedom.

This would be a freedom à la Spinoza—a freedom that is "critical." Critical freedom, wrote James Tully in *Strange Multiplicity*, is the practice of seeing the specificity of one's own world as one among others. Inter-temporally, this freedom sees the present as fully engaged with a past that, insofar as it is part of the present, can be re-written a little more freely. The act of inter-facing can do that. The term, or illusion, of universality may not be the most felicitous one to characterize this act, but accompanied by the verb "confronting" it makes sense beyond a relativism that implies turning one's back on such faces.

NOTES

1. Doris Salcedo, *Shibboleth*, 65.
2. Doris Salcedo's Unilever Commission *Shibboleth* at the Tate Modern in London consisted of a long, deep, and elaborate crack in the floor of the Turbine Hall. The catalogue explains the artist's attempt to put the global division between people down literally. The term "infra-humanity" must be understood in that context.
3. Making videos is an attempt to respond creatively to the constant actuality of the topics of the contemporary. By definition, the contemporary causes the belatedness of publications. This will always make it impossible to limit our research to libraries. Videos, while of course not full accounts either, preserve something of the voice of the subjects they stage.
4. Jan van der Lubbe and Aart van Zoest, "Subtypes of Inference and Their Relevance for Artificial Intelligence," 805.
5. Ibid., 806.
6. For an overview of the term "abduction" and the sources in Peirce's oeuvre, see Harry G. Frankfurt, "Peirce's Notion of Abduction."
7. For a complete list or for stills and synopses of these works, see http://www.miekebal.org.
8. Mieke Bal, *Nothing Is Missing,* multiple-channel video installation. See also Mieke Bal, "Nothing Is Missing," *Intermédialités* 8.
9. Alexander Gottlieb Baumgarten, *Aesthetica.*
10. Jacques Rancière, *Disagreement.*
11. Chantal Mouffe, *On the Political*, 9.
12. Anthony Uhlmann, "Image and Intuition in Beckett's *Film.*"
13. Jean-Joseph Goux, *Oedipe philosophe.*
14. I have been stimulated to think in terms of second-personhood by Lorraine Code, *What Can She Know?* and her later work *Rhetorical Spaces.* Louise Antony, in "'Human Nature' and Its Role in Feminist Theory," develops a universalist definition of humanity from a feminist perspective in order to avoid the universal-relativist trap.
15. Gayatri Chakravorty Spivak's productive term in *A Critique of Postcolonial Reason.*
16. My argument here is based on the ideas put forth by Henri Bergson in *Time and Free Will*; *Creative Evolution*; and *Matter and Memory.* Gilles Deleuze has devoted a book-length study to Bergson: *Bergsonism.*
17. Moira Gatens and Genevieve Lloyd, *Collective Imaginings.*
18. Mieke Bal, *Quoting Caravaggio.*
19. Richard Brilliant, *Portraiture*, 9.
20. "Occasionality" refers to the reality depicted; in the case of the portrait, it refers to the sitter. See Hans Georg Gadamer, *Truth and Method.*
21. Brilliant, *Portraiture*, 8.
22. Ibid., 10, emphasis added.
23. Johannes Fabian, *Anthropology with an Attitude.*
24. The notion of "giving time" is a reference to Jacques Derrida, *Given Time.*

25. Beckett writes to Alan Schneider here, in Samuel Beckett and Alan Schneider, *No Author Better Served*, as quoted in Anthony Uhlmann, "Image and Intuition," 101–2. For more background on Beckett's *Film*, see Samuel Beckett, *Film: Complete Scenario, Illustrations, Production Shots.*

26. Uhlmann, "Image and Intuition," 103.

27. Gatens and Lloyd, *Collective Imaginings*, 40.

28. Paola Marrati, *Gilles Deleuze*, 48; my translation.

29. Gatens and Lloyd, *Collective Imaginings*, 52.

30. Marianna Torgovnick, "The Politics of the 'We,'" 265.

31. Gatens and Lloyd, *Collective Imaginings*, 81.

32. Ibid., 82.

33. Hamid Naficy, *An Accented Cinema.*

34. E.g., Michel Chion, *Un art sonore, le cinéma.*

35. On the rhetoric of sincerity, see Ernst van Alphen, Mieke Bal, and Carel Smith, eds., *The Rhetoric of Sincerity.*

36. The phrase "present pasts" alludes to the title of Andreas Huyssen's book, much of which is relevant to the present discussion: Andreas Huyssen, *Present Pasts.*

37. Gilles Deleuze, *Cinema 1*, 99.

BIBLIOGRAPHY

Antony, Louise M. "'Human Nature' and Its Role in Feminist Theory." In Janet A. Kourany, ed., *Philosophy in a Feminist Voice: Critiques and Reconstructions*, 63–91. Princeton: Princeton University Press, 1998.

Bal, Mieke. "Nothing Is Missing." *Intermédialités* 8 (Autumn 2006): 189–224.

——. *Nothing Is Missing,* multiple-channel video installation. DVD, furniture, and mixed media, 4 to 15 units, 28 to 35 minutes, looped.

——. *Quoting Caravaggio: Contemporary Art, Preposterous History.* Chicago: University of Chicago Press, 1999.

——. *Travelling Concepts in the Humanities: A Rough Guide.* Toronto: University of Toronto Press, 2002.

Baumgarten, Alexander Gottlieb. *Aesthetica.* Hildesheim, Germany: Olms, 1970 [1750].

Beckett, Samuel. *Film: Complete Scenario, Illustrations, Production Shots.* New York: Grove, 1969.

——, and Alan Schneider. *No Author Better Served: The Correspondence of Samuel Beckett and Alan Schneider.* Ed. Maurice Harmon. Cambridge: Harvard University Press, 1998.

Bergson, Henri. *Creative Evolution.* Trans. A. Mitchell. Landham, MD: University Press of America, 1983 [1907].

——. *Matter and Memory.* Trans. N. M. Paul and W. S. Palmer. New York: Zone Books, 1991 [1896].

——. *Time and Free Will: An Essay on the Immediate Data of Consciousness.* Trans. F. L. Pogson. New York: Harper and Row, 1960 [1889].

Brilliant, Richard. *Portraiture.* Cambridge: Harvard University Press, 1991.

Chion, Michel. *Un art sonore, le cinéma: Histoire, esthétique, poétique.* Paris: Cahiers du cinéma, 2003.

Code, Lorraine. *Rhetorical Spaces: Essays on Gendered Location.* New York and London: Routledge, 1995.

——. *What Can She Know? Feminist Epistemology and the Construction of Knowledge.* Ithaca: Cornell University Press, 1991.

Deleuze, Gilles. *Bergsonism.* Trans. Hugh Tomlinson and Barbara Habberjam. New York: Zone Books, 1988.

——. *Cinema 1: The Movement-Image.* Trans. Hugh Tomlinson and Barbara Habberjam. London: Athlone Press, 1986.

Derrida, Jacques. *Given Time: I. Counterfeit Money.* Trans. Peggy Kamuf. Chicago: University of Chicago Press, 1992.

Fabian, Johannes. *Anthropology with an Attitude: Critical Essays.* Stanford: Stanford University Press, 2001.

Frankfurt, Harry G. "Peirce's Notion of Abduction." *The Journal of Philosophy* 55, no. 14 (July 1958): 593–97.

Gadamer, Hans Georg. *Truth and Method.* Trans. Joel Weinsheimer and Donald G. Marshall. 2nd rev. ed. New York: Crossroad, 1989.

Gatens, Moira, and Genevieve Lloyd. *Collective Imaginings: Spinoza, Past and Present.* New York and London: Routledge, 1999.

Goux, Jean-Joseph. *Oedipe philosophe.* Paris: Aubier, 1990.

Huyssen, Andreas. *Present Pasts: Urban Palimpsests and the Politics of Memory.* Stanford: Stanford University Press, 2003.

Marrati, Paola. *Gilles Deleuze: Cinéma et philosophie.* Paris: PUF, 2003.

Mouffe, Chantal. *On the Political.* New York and London: Routledge, 2005.

Naficy, Hamid. *An Accented Cinema: Exilic and Diasporic Filmmaking.* Princeton: Princeton University Press, 2001.

Rancière, Jacques. *Disagreement: Politics and Philosophy.* Trans. Julie Rose. Minneapolis: University of Minnesota Press, 1999.

Salcedo, Doris. *Shibboleth.* Exhibition catalogue. London: Tate, 2007.

Spivak, Gayatri Chakravorty. *A Critique of Postcolonial Reason: Toward a History of the Vanishing Present.* Cambridge: Harvard University Press, 1999.

Torgovnick, Marianna. "The Politics of the 'We.'" In Marianna Torgovnick, ed., *Eloquent Obsessions: Writing Cultural Criticism*, 260–78. Durham: Duke University Press, 1994.

Tully, James. *Strange Multiplicity: Constitutionalism in an Age of Diversity.* Cambridge and New York: Cambridge University Press, 1995.

Uhlmann, Anthony. "Image and Intuition in Beckett's *Film.*" *SubStance* 33, no. 2 (2004): 90–106.

Van Alphen, Ernst, Mieke Bal, and Carel Smith, eds. *The Rhetoric of Sincerity*. Stanford: Stanford University Press, 2008.

Van der Lubbe, Jan, and Aart van Zoest. "Subtypes of Inference and Their Relevance for Artificial Intelligence." In Irmengard Rauch and Gerald F. Carr, eds., *Semiotics Around the World: Synthesis in Diversity*, 805–8. Berlin and New York: Mouton de Gruyter, 1997.

6

OBJECTS OF RETURN

Marianne Hirsch

Edek resumed his digging. He dug and he dug. Half of the outhouse's foundation now seemed to be exposed. Edek got down on his knees, and dug a hole at the base of the foundations. Suddenly he stiffened.

"I think I did find something." Everyone crowded in. . . .

He reached under the foundation and dug around with his fingers. He was lying stretched out on the ground.

"I got it," Edek said breathlessly. He pulled out a small object, and began removing the dirt from its surface. The old man and woman tried to get closer.

"What has he got? What has he got?" the old woman said. . . .

Edek got up. He had cleaned up the object. Ruth could see it. It was a small, rusty, flat tin. "I did find it," Edek said, and smiled.[1]

INCONGRUOUS OBJECTS

At the end of Lily Brett's 1999 novel, *Too Many Men*, Edek and his Australian-born daughter Ruth return one more time to Kamedulska Street in Lódz where Edek had grown up as a small boy and young man in the 1920s and 1930s. They had already been there several times and, each time, had discovered additional objects that provided clues to Edek's and his family's past. Ruth had gone there by herself to buy, for inordinate sums of money, her grandmother's tea service and other personal items that the old couple living in Edek's former apartment brought out for her in a slow and emotionally tortuous process of extortion. But after traveling on from Lódz to Kraków and then to Auschwitz, where Edek and his wife Rooshka had survived the war, Edek insisted on returning to Lódz and to Kamedulska Street once more to retrieve an additional item of immense personal value. "Did they find gold?" the neighbors kept asking, but the old couple had already searched every inch of ground and come up empty. To his great joy, Edek does find the precious object buried in the ground: it is "a small, rusty, flat tin."

It isn't until later, at their hotel, that Edek opens the small tin. Ruth "could feel the dread in her mouth, in her throat, in her lungs, and in her stomach. . . . The tin held only one thing. Edek removed the object from the tin. It was a photograph. A small photograph. . . . It was a photograph of her mother. . . . Rooshka was holding a small baby. The small baby was Ruth. . . . 'It does look like you,' Edek said. 'But it is not you.' Ruth felt sick" (518). Edek then proceeds to tell Ruth a story concerning her parents that she had never heard before. After liberation, Edek and Rooshka had found each other again and had had a baby boy in the German displaced persons camp Feldafing. The baby had been born with a heart problem that required a kind of care that these stateless Auschwitz survivors were not in a position to provide. At the advice of their doctor, they made the excruciating decision to give him up to a wealthy German couple for adoption. Before giving him away, Edek had taken a photograph of the baby. Rooshka, however, "'was angry. She did say that if we are going to give him away, he will be out of our lives, so why should we pretend with a photograph that he is part of us. . . . Mum did tell me to throw away the photograph. But I did not want to throw it away'" (524). Edek gave the photo to his cousin Herschel, who was going back to Kamedulska Street, which, he believed, "'was still more his home than the barracks.'" Herschel took the photo with him and, discovering that this could, in fact, never again be his home, buried it in the yard under the outhouse before returning to the DP camp.

Too Many Men belongs to a genre of Holocaust narrative that has been increasingly prevalent in recent years: the *narrative of return*, in which a Holocaust survivor, accompanied by an adult child, returns to his or her former home in Eastern Europe, or in which children of survivors return to find their parents' former homes, to "walk where they once walked."[2] Memoirs by these children of survivors dominate this narrative genre, but *Too Many Men* provides a rich fictional example and the chance to discuss the characteristics of return plots that are generally punctuated by images and objects that mediate acts of return.

Narratives of return are quest plots holding out, and forever frustrating, the promise of revelation and recovery; thus Edek's discovery of the metal tin and the baby's photograph offers a rare epiphanic instant in this genre. And yet, characteristically perhaps, this moment of disclosure and satisfaction serves only to raise another set of questions that defer any possibility of narrative closure. Why, when Edek's baby was born after liberation, in Germany, was his photo taken back to Lódz, to Kamedulska Street, to be buried there? And why does Edek spend such enormous sums of money and effort to go back to his former home one more time to search for and to retrieve the photograph? If the photo is so important to him, why does he not dig for it on his first visit; why, in fact, does he wait? Narratives of return, like *Too Many Men*, abound in implausible plot details such as these. What, actually, does Ruth find out about her parents and about herself when her father succeeds in unearthing the photo of her lost brother? What can these moments of narrative fracture and incongruity tell us about the needs and impulses that engender return in different generations and about the scenarios of intergenerational transmission performed in and by acts of return?[3]

In this essay, I read Brett's novel alongside two other works that clarify the incongruities, the implausibilities and impossibilities, and the fractured shapes characterizing the impulse to return and its narrative and visual enactments: Palestinian writer Ghassan Kanafani's 1969 novella *Return to Haifa*, a work that deals not with the Holocaust but with the Nakba, the Palestinian expulsions, literally "catastrophe" of 1948, and the Eurydice series of Bracha Lichtenberg-Ettinger, an Israeli visual artist who is the daughter of Holocaust survivors.[4] These three works enable us to look, in particular, at the role that objects (photographs, domestic interiors, household objects, items of clothing) play in return stories, marking their sites of implausibility and incommensurability. Such testimonial objects, lost and again found, structure plots of return: they can embody memory and thus trigger affect shared across generations. But as heavily symbolic and overdetermined sites of contestation, they can also mediate the political, economic, and juridical claims of dispossession and recovery that often motivate return stories.

Read together, these three works stage the impulse to return as a fractured encounter between generations, between cultures, and between mutually imbricated histories occurring in a layered present. From Australia, New York, and Israel to Poland and back, from the West Bank to Haifa, from a layered present to a complicated past, return is desired as much as it is impossible. In focusing on the emblematic figure of the lost child, however, these works expose the deepest layers of the contradictory psychology of return and the depths of dispossession that reach beyond specific historical circumstances. How can divergent histories that expose children to danger and abandonment be thought together without flattening or blurring the differences between them? Perhaps in a feminist, *connective* rather than *comparative*, reading that moves between global and intimate concerns by attending precisely to the intimate details that animate each case even while enabling the discovery of shared motivations and shared tropes.[5] Such a feminist reading, as I see it, pays attention to the political dimensions of the familial and domestic, and to the gender and power dynamics of contested histories. It foregrounds affect and embodiment and a concern for justice and acts of repair.

In Kanafani's text, a Palestinian couple drives from Ramallah to the house in Haifa that they were forced to leave in 1948. It is June 1967, twenty years later, and Said S. and Safiya join many of their neighbors and friends curious to revisit the homes they had left behind and that they were allowed to visit after the Israeli annexation of the West Bank and the opening of the borders. As they approach Haifa, Said S. "felt sorrow mounting from inside him. . . . No, the memory did not come back to him little by little, but filled the whole inside of his head, as the walls of stone collapsed and piled on top of each other. Things and events came suddenly, beginning to disintegrate and filling his body."[6] Return to place literally loosens the defensive walls against the sorrow of loss that refugees build up over decades and that they pass down to their children. Just as Ruth responds to the photograph in *Too Many Men*, so Said S. and Safiya respond viscerally, with trembling, tears, sweat, and overpowering physical feelings of torment. As the couple approaches their former house, the streets they cross, the smells of the landscape, the topography of the city all trigger bodily responses that are not exactly memories, but reenactments and reincarnations of the events of the day in 1948 when they left their home. The past overpowers the present, "suddenly, cutting like a knife" (102), and we are with Said S. in 1948 as he desperately attempts to get back to his wife through the bullets and confusion on the city streets; we also see her haste to get to him, and her inability to fight her way back through the flood of refugees to the house where her baby, Khaldun, remains asleep in his crib, tragically left behind. Later that day, offshore, as the boats took them away from Haifa, "they were incapable of feeling anything" (107). The loss is so overwhelming that for twenty years, Khaldun remains a family secret: Khaldun's name is rarely pronounced in their house, and then only in a whisper. Their two younger children do not know about their lost brother. And even as they drive toward Haifa together in 1967, neither Said S. nor Safiya, who talk about everything else during the journey, "had uttered a syllable about the matter that had brought them there" (100). On the surface, the trip is about seeing their house again—as they say, "just to see it" (108).

Both these fictional works represent refugees' and exiles' re-encounter with the material textures of their daily lives in the past.[7] "Habit," Paul Connerton writes in his book *How Societies Remember*, "is a knowledge and a remembering in the hands and in the body, and in the cultivation of habit it is our body which 'understands.'"[8] In returning to the spaces and objects of the past, displaced people can remember the embodied practices and the incorporated knowledge that they associate with home. When Said S. slows his car "before reaching the turn which he knew to be hidden behind the foot of the hill,"[9] when he "looked at all the little things which he knew would frighten him or make him lose his balance: the bell, the copper door knocker, the pencil scribblings on the wall, the electricity box, the four steps broken in the middle, the fine curved railing which your hand slipped along," he reanimates deep habits and sense memories. Ordinary objects mediate the memory of returnees through the particular embodied practices that they re-elicit. And these embodied practices can also revive the affect of the past, overlaid with the shadows of loss and dispossession.

Said and Safiya notice every detail of the house, comparing and contrasting the present with the past "like someone who had just awoken from a long period of unconsciousness" (112). Much remained exactly the same: the picture of Jerusalem on one wall, the small Persian carpet on the other. The glass vase on the table had been replaced with a wooden one, but the peacock feathers inside it were still the same, though of the seven feathers that had been there, only five remained. Both wanted to know what happened to the other two.

Somehow, those two missing feathers become signifiers of the incommensurability of return—a measure of the time that had passed and the life lived by other people and other bodies in the same space and among the same objects. Emerging from this bodily re-immersion in his former home, Said S. begins to realize that, for years, other feet have shuffled down the long hallway, and others have eaten at his table: "How very strange! Three pairs of eyes all looking at the same things . . . and how differently everyone sees them!" (113). In Kanafani's novella, the third pair of eyes belong to Miriam, the wife of the deceased Evrat Kushen, both Holocaust survivors who had been given the house by the Jewish Agency only a few days after Said S. and Safiya left it.[10] "And with the house, he was given a child, five months old!" (120), Miriam tells Said S. and Safiya as they sit in the living room that all three of them consider their own. Miriam also tells them how she had been ready to return to the Italian DP camp to which they had been sent after the war because of the disturbing scene she had witnessed during those days in 1948: Jewish soldiers throwing a dead Arab child, covered with blood, into a truck "as if he were a piece of wood." When they adopted the baby, Evrat Kushen hoped that his wife would be able to heal from the shock of that vision.

As Said S. and Safiya discuss whether to wait for the return of Khaldun, who had been raised by his Jewish parents as Dov, or whether to leave immediately, accepting the fact that their son had irrevocably been taken from them, they surprisingly equate their child with the house that had been, and was now no longer, theirs. Both house and child are invested with agency and power—to accept or to deny their former owners/parents. As Said tells Safiya, "'Don't you have those same awful feelings which came over me while I was driving the car through the streets of Haifa? I felt that I knew it and that it had denied me. The same feelings came over me when I was in the house here. This is our house. Can you imagine that? Can you imagine that it would deny us?'" (123). What could figure the enormity of their dispossession as powerfully as the loss of a child, or a child's refusal to recognize his parents? When Khaldun/Dov finally appears on the scene, he is wearing an Israeli uniform.

Although, in Kanafani's novella, the lost child structures the story of return, the novella's plot does not fully motivate the loss of Khaldun: we are told that Safiya tried, desperately, to return to the house to fetch her baby, but how, we cannot help but wonder, could she have left him there in the first place? This narrative implausibility is compounded by other textual incongruities, most notably Dov's revelation that "they" (his parents) only told him "three or four years ago" (131) that he was not their biological son, even though he said earlier that "my father was killed in Sinai eleven years ago" (129).[11] Is this temporal disjunction a mistake on Kanafani's part, or an indication of the son's very belated acceptance of his adoption? These implausible elements of the story and the questions they raise produce moments of fracture in which different plot possibilities are overlaid on one another with no possible resolution. They can be motivated only on the level of fantasy and symbol—as the measures of a failed maternity and paternity in a time of historical extremity, and as emblems of the radical dispossession that is the result of Israeli occupation.

For Said S. and Safiya, as for Edek and Rooshka in *Too Many Men*, the lost child remains a shameful and well-kept secret, haunting and layering the present. When Edek regrets giving his baby up for adoption and worries that he had made the wrong decision, Ruth attempts to alleviate her father's guilt by insisting that "You did nothing wrong."[12] Here, also, extreme historical circumstances fracture family life and disable parental nurturance. Just as Said S. and Safiya need to return to their former home and to reencounter the objects that trigger body memories and with them the emotions of inconsolable loss they had so long suppressed, so Edek needs to find the photograph of his baby son if he is to tell the story to his daughter. More than objects of intimacy—Ruth's grandmother's tea service, or her grandfather's overcoat and the photographs that are in one of its pockets—the photo of the lost child figures the expulsion from home and the impossibility of return. In the narrative, we need to wait for its revelatory power; we need to witness the progressive discovery Edek and, with him, Ruth undergo. Suspense, partial disclosures, and delayed revelations structure the plot: several scenes of digging have to precede the unearthing of the small tin can. In both texts, the loss of the child, associated with guilt and shame, is deeply suppressed. It can be brought into the open and confronted only gradually, by crossing immense temporal and spatial divides.

For Edek, the photograph becomes the medium of a narrative shared across generations. Ruth wonders why her parents had never told her about her baby brother and insists that this is her story as much as it is her mother's and father's: "It was impossible to grow up unaffected. The things that happened to you and to Mum became part of my life. Not the original experiences, but the effects of the experiences'" (527). It is these effects that motivate Ruth's journey to Poland, her need to imagine her parents' lives, her tireless search for every object and every detail of their past. They motivate her repeated returns to Kamedulska Street, and her need to go there with her father. And, through a process of unconscious transmission of affect, they motivate a recurrent nightmare that plagues Ruth throughout her trip to Poland, before she ever sees the photo or learns about her lost brother:

> She had had one of her recurring nightmares. The worst one, the one in which she was a mother. The children were almost always babies. . . . In these dreams she lost her babies or starved them. She misplaced them. Left them on trains or buses. . . . The abandonment in her dreams was never intentional. She simply forgot that she had given birth to and brought home a baby. When in her dreams she realized what she had done, she was mortified (113).

How does the act of returning to place and how do the objects found there inflect the process of affective transmission that so profoundly shapes the postmemory of children of exiles and refugees?

RETURNING BODIES

In her latest book on memory, *Der lange Schatten der Vergangenheit*, the German critic Aleida Assmann reflects on the role of objects and places as triggers of body or sense memory.[13] Invoking the German reflexive formulation of "ich erinnere mich" she distinguishes what she calls the verbal and declarative, active "ich-Gedächtnis" (I-memory) from the more passive "mich-Gedächtnis" (mememory), appealing to the body and the senses rather than to language or reason.[14]

Assmann's "mich-Gedächtnis" is the site of involuntary memory that is often activated and mediated by the encounter with objects and places from the past. Scholars of memory sites like James Young and Andreas Huyssen have been skeptical of what they deem a Romantic notion that endows objects and places with aura or with memory. In response, Assmann specifies that, though objects and places do not themselves carry qualities of past lives, they do hold whatever we ourselves project onto them, or invest them with. When we leave them behind we bring something of that investment along, but part of it also remains there, embedded in the object or the place itself. Assmann uses the metaphor of the classical Greek legal concept of the *symbolon* to conceptualize this. To draw up a legal contract, a symbolic object was broken in half, and one of those halves was given to each of the parties involved. When the two parties brought the two halves together at a future time, and they fit, their identity and the legal force of the contract could be ratified. Return journeys *can* have the *effect* of such a re-connection of severed parts, and if this indeed happens, they can release latent, repressed, or dissociated memories—memories that, metaphorically speaking, remained behind, as it were, concealed within the object. And, in so doing, they can cause them to surface and become re-embodied. Objects and places, there-fore, Assmann argues, can function as triggers of remembrance that connect us, bodily and thus also emotionally, with the object world we inhabit.[15] In her formulation, the "mich-Gedächtnis" functions as a system of potential reso-nances, of chords that, in the right circumstances—during journeys of return, for example—can be made to reverberate.

But can the metaphor of the *symbolon* cover cases of massive historic frac-tures, such as the ones introduced by the Holocaust and the Nakba? Would not contracts lose their legal force in such cases, so much so that the pieces would no longer be expected to fit together again? Worn away not only by time but also by a traumatic history of displacement, forgetting, and erasure, places change and objects are used by other, perhaps hostile, owners, over time coming to merely approximate the spaces and objects that were left behind. Cups and plates chip, peacock feathers disappear, wooden vases replace glass ones, keys to houses, obses-sively kept in exile, no longer open doors. "Home" becomes a place of no return.

And yet, embodied journeys of return, corporeal encounters with place, do have the capacity to create sparks of connection that activate remembrance and thus reactivate the trauma of loss. In the register of the more passive "mich-Gedächtnis," or what Diana Taylor terms the "repertoire,"[16] they may not re-lease full accounts of the past, but they can bring back its gestures and its affects. Perhaps the sparks created when the two parts of a severed power line touch ever so briefly constitute a more apt image for this than the ancient *symbolon* cut in half. The intense bodily responses to the visits of return that we see in both these novels testify to the power of these sparks of reconnection that increase expectation and thus also intensify frustration.

The powerful body memory engendered in return is compounded in these narratives by the trope of the lost child that clarifies the enormity of the stakes involved. In contrast to Brett and Kanafani, W. G. Sebald's *Austerlitz*, published in 2001, another paradigmatic narrative of return, reverses the generations by staging the return of the lost amnesiac child in search of his parents and his own past self.[17] Here too bodily symptoms signal found sites and engender moments of reconnection, often without cognitive recognition: "I could tell, by the prickling of my scalp," Austerlitz says of a scene that was "brought back out of my past" (151). "It was as if I had already been this way before and memories were revealing themselves to me not by means of any mental effort but through my senses" (150). When he is handed the photo of the little page boy and is told, "This is you, Jacquot" (183), Austerlitz is "speechless and uncomprehending, incapable of any lucid thought" (184). "I could not imagine who or what I was" (185), he repeats. In the face of expulsion and expropriation—especially childhood expulsion—home and identity are in themselves implausible and objects remain alienating and strange.

GENERATIONS AND SURROGATIONS

The impossibility and implausibility of return is intensified if descendants who were never there earlier return to the sites of trauma. Can they even attempt to put the pieces together, to create the spark? Or is the point of connection and the physical contact with objects lost with the survivor generation? What if several generations pass? What if traces are deliberately erased and forgetting is imposed on those who are abducted or expelled, as Saidiya Hartman asks in her moving memoir of "return" to the slave routes of Ghana, tellingly entitled *Lose Your Mother*?[18] Her narrative, like other second and subsequent generation stories of return, attempts to reclaim memory and connection to the objects and places of the past, even while making evident the irreparability of the breach. Narrative incongruity in fictional accounts may well serve the purpose of signaling the fractures and implausibities underlying both home and return in the accounts of postmemorial generations that have inherited the loss. And images, objects, and places function as sites where these implausibilities manifest themselves.

In *Too Many Men*, Ruth says to Edek that "so much of what happened in your life became part of my life."[19] Along with stories, behaviors, and symptoms, parents do transmit to their children aspects of their relationship to places and objects from the past. Ruth had wanted to visit Poland the first time "just to see that her mother and father came from somewhere. To see the bricks and the mortar. The second time was an attempt to be less overwhelmed than she was the first time. To try and not to cry all day and night. . . . And now she was here to walk on this earth with her father."[20] He identifies places and objects, gives her information, but also, together, they are able to relive the most difficult and painful moments of his past—to transmit and to receive the sparks of reconnection. These are often provocative and disturbing, as when Said S. and Safiya confront Dov with the reality of his double identity, or when Ruth cannot evade the running mental dialogue with Rudolf Hoess, the commandant of Auschwitz whose voice, implausibly again, addresses and argues with her as soon as she arrives in Poland, from the dead, or, as he presents it, from "Zweites Himmel Lager" second heaven camp, as part of his own rehabilitation program.

This need on the part of the child born after the war, and after the moment of expulsion and expropriation, to visit the places from which her parents were evicted provides another explanation for cousin Herschel's return and the burial of the baby's photograph in Lódz in *Too Many Men*. It comes from a yearning to find a world before the loss has occurred, before the Rudolf Hoesses dominate the scene—from a need for an irrecoverable lost innocence that descendants of survivors imagine and project.

With the small tin and the photo inside it, Edek is unearthing more than his own repressed feelings of loss. He is demonstrating to Ruth her surrogate role: she is not the first child to "return" "home" to Lódz; the baby had already come back before she did, albeit in the guise of a photo. As Joseph Roach argues in *Cities of the Dead*, cultural memory of loss works through a genealogical network of relations we might think of as surrogation: memory is repetition but always with some change, reincarnation but with a difference.[21] Those of us living in the present do not take the place of the dead but live among or alongside them. In encountering the baby's photo, Ruth comes to understand her role as surrogate. The lost brother's photo is literally dug up from under the foundation of the house. His image, like and also unlike her, emerges as the fantasmatic figure shared by all children of survivors who tend to think of themselves as "memorial candles"—stand-ins for another lost child, who becomes responsible for perpetuating remembrance, for combating forgetting, for speaking in two overlapping voices.[22]

The structure of surrogation functions even more literally in Kanafani's text. The children Said S. and Safiya have after losing Khaldun are called Khalid and Khalida. These children do not know about their brother. And yet, on some level, they may have learned that, in the familial economy of loss, they are taking his place. At the end of the novel, Said emotionally authorizes Khalid to take up arms for Palestine, to win back the home from which they were evicted. Mythically, brother will fight against brother.

Connecting these two texts, however, shows how differently memory can function in the different contexts in which journeys of return take place. When Edek returns to Kamedulska Street, the old couple residing there worries that he plans to reclaim his property. Nothing could be further from his intentions: Edek and Ruth are there to find the past, not the present. Even though Edek enjoys the food of his youth and feels at home in his language, he is uninterested in Poland, cannot wait to leave it. Ruth could not imagine living there: her project is one of mourning and psychic repair, not recuperation. Bringing her grandmother's tea service to New York promises to reconnect some of the disparate parts of her life, to achieve continuity with a severed past—not to bring it into the present. But Hoess's constant whisper, overlaid onto her musings about her parents' and grandparents' world of before, shows her how much the past and also the future are dominated and overshadowed by the incontrovertible fact of the genocide. No revelation or recovery can heal the breach.

In Kanafani's text, we see an entirely different economy at work. "'You can stay for a while in our house,'" Said S. says to Miriam and to Dov as he leaves. It is *his* house, and he imagines that his son Khalid will help recover it. Khalid had wanted to join the *fedayeen*, to become a guerilla and sacrifice his life for the struggle, but his father and mother had been opposed. Now, driving back to Ramallah, Said hopes that his son has left while they were away. Memory serves a future of armed struggle and resistance here, not one of mourning or melancholy. And, in a context in which the conflict continues and resolution cannot yet be envisioned, return serves the cause of legal and moral claims of recovery.

But the fantasmatic structure of surrogation functions in a more disturbing and open-ended fashion as well. In both texts, the lost child's survival is, for a while at least, submerged in ambiguity. Said S. and Safiya do not know whether they will find Khaldun when they return to Haifa. Ruth harbors fantasies of searching for and finding her lost brother, and, in fact, the novel, implausibly again, holds out that possibility: a German woman Ruth meets on her trip tells her about a young Christian German man named Gerhard who looks exactly like her and who, though gentile, identifies profoundly with Jews. At the end of the novel, Ruth, in a fantasy of recovering her lost brother, sets out to search for Gerhard. Could the lost child, then, function in these texts along the logic of the uncanny—as the embodiment of childhood innocence and hope, of the belief in a future, irrevocably lost with war and dispossession? In this schema, surrogation would work backward toward a primordial past, rather than forward into the future, and Edek, as well as Said S., would be finding not their sons but their own childhood selves—lost, unprotected, neglected, forgotten, repressed, but returning, perpetually and uncannily, to haunt a tainted present and to hold out the vision of an alternative ethical and affective structure.

VISUAL RETURNS

I turn now to a third body of works that exhibit a visual aesthetics of return characterized by fracture, overlay, and superimposition. The works by Bracha Lichtenberg-Ettinger, a second-generation Lacanian psychoanalyst and feminist visual artist allow us to measure the political and psychic implications of the repetitions and irresolutions of return.

In her Eurydice series, produced between 1990 and 2001, Ettinger goes back to a street in Lódz at a moment "before" the Holocaust.

A 1937 street photograph of Ettinger's parents from the Polish city of Lódz has become an obsessive image that recurs throughout her visual work in many iterations (figure 6.1). In the image, her young parents are walking smilingly and energetically down a street in their town, exuding comfort and safety. They are happy to look and be looked at, to display and perform their sense of belonging in this city and its urban spaces. In a label on her Web site, the artist informs viewers that, unlike her parents, the friend walking on the street with them was later killed by the Nazis. The image of her young parents taken before her own birth appears in Ettinger's *Mamalangue* superimposed onto another image, a washed-out photograph of her own childhood face (figure 6.2).

The child's smile is covered, almost erased, by the mother's smiling figure. The prewar stroll through the city, the couple walking toward a future they could not yet imagine, bleeds into the face of the child who grew up in a distant place, dominated by stories and histories that preceded her birth. Here, in overlay fashion, are the past and the present, two worlds that the postwar child longs to bridge: the world her parents once knew—where the Holocaust had not yet happened—and her own world, "after Auschwitz." In projecting her own face onto and into the spaces of the past, Ettinger absorbs some of the embodied

6.1 Bracha Lichtenberg-Ettinger, *Street photo, Lódz, 1937*

6.2 Bracha Lichtenberg-Ettinger, *Mamalangue—Borderline conditions and pathological narcissism*, no. 5

practices of that past moment, enacting a kind of return journey in photographic mode. This journey is characterized by structures of superimposition and overlay similar to those in Brett's and Kanafani's fictions. Past and present coexist in layered fashion, and their interaction is dominated by objects that provoke deep body memory and the affects it triggers.

Thus, in Ettinger's composite images, the Lódz street photo of her parents, and the image of her childhood face, are often juxtaposed, overlaid, or blended with yet a third image—a disturbing, well-known photograph of a group of naked Jewish women, some holding children, herded to their execution by Einsatzgruppen in Poland (figures 6.3 and 6.4). This image, no doubt taken by one of the Nazi photographers who accompanied the Einsatzgruppen, serves as the basis for Ettinger's Eurydice series. This is not a space to which one would want to return; it is the antithesis of "home."

For this artwork, Ettinger made reproductions of various details of the Einsatzgruppen photo, which she then ran through a photocopier, enlarged, cut into strips, mounted on walls, and tinted with India ink and purple paint to a point where all details are washed out or made virtually invisible. The juxtapositions of the Lódz and Einsatzgruppen photos—one taken by a prewar street photographer, a fellow citizen, the other snapped and shaped by the annihilating Nazi gaze conflating the camera with the gun—illustrate the child's deep fear of parental impotence in a time of extremity that dominates all the texts under discussion. They illustrate the underside of return, the fear that violence will be repeated, that, as in Eurydice's backward look, return will prove to be deadly. As objects coming out of the past, the superimposed images contain and activate those fears, and Ettinger mobilizes them in her superimpositions.

The women walking toward their death, holding their babies, suffer the ultimate failure of parental care: they cannot protect their children, or themselves, from annihilation. They are witnesses and victims of the ultimate breach of a social contract in which adults are supposed to protect infants, rather than murdering them. In these composite images, the artist, as child, becomes the surrogate of the dead baby; the infant held by the mother in the picture becomes, by implication, her own phantom sibling or her fragile child self.

In the Eurydice images, the photos from "before" cannot be separated from the photos taken "during" and "after" the destruction. As a child born after the war—a child who might easily never have been born at all had her parents' fates taken a slightly different turn—Ettinger is unable to return to the spaces from "before" without the superimposed, layered, screen image of the atrocity "during." The prewar photo from the family album—from the seemingly protected intimate and embodied space of the family and its repertoires—cannot be insulated from the collective, anonymous images in the killing fields, and the child born after the war is inevitably haunted by the phantom sibling. The two kinds of images, and the three temporalities, are inextricably linked. In this way, Ettinger's juxtapositions forge a powerful antinostalgic idiom for the postmemorial subject. Return, even metaphoric return, cannot jump over the breach of expulsion, expropriation, and murder.

In foregrounding and recasting Eurydice into the maternal figure who lost her child, Ettinger, moreover, is reframing the father/son and father/daughter perspectives presented in Kanafani's and Brett's novels. Ettinger's Eurydice is the maternal figure who returns from Hades having witnessed the loss of her child. In

6.3 Bracha Lichtenberg-Ettinger, *Eurydice*, no. 37

6.4 Bracha Lichtenberg-Ettinger, *Eurydice*, no. 5

her powerful reading of the Eurydice series, Griselda Pollock sees Eurydice as the woman precariously alive "between two deaths:" for the women in the ravine, this is the brief moment between being shot by the camera and by the gun. Eurydice's story is no doubt the prototypical narrative of impossible return. For Pollock, however, Ettinger, the daughter artist, is reframing the Orphic gaze of no return, in favor of an aesthetic of what Ettinger calls "wit(h)ness" and "co-affectivity."[23] In subjecting her original images to technologies of mechanical reproduction, in degrading, recycling, reproducing, and painting over them, Ettinger underscores the distance and anonymity of the camera gaze. But, at the same time, she allows all of these images to invade, inhabit, and haunt her, and she therefore inscribes them with her own very invested act of looking, exposing, in the images, her own needs and desires—her own fears and nightmares. In Pollock's reading, the purple paint is a physical touch that marks the images with "the color of grief."[24]

But more still is at stake in a few of the images from this series that include yet one more level of superimposition (figure 6.5).

The grid in this image is of a World War I map of Palestine, and of aerial views of Palestinian spaces taken by German war planes during World War I when Palestine was occupied by the British. Ettinger, born in Israel after World War II, has inherited other traces of loss—traces of Palestinian spaces that she maps onto the spaces of Polish streets before World War II.

These competing spaces and temporalities become part of a multilayered psychic grid, unconsciously transmitted, merging geography and history, and challenging any clear chronology or topography of return. The views are aerial shots taken by military airplanes. In including them in her own family image, Ettinger dramatizes, in a closer and more intimate manner, the irreconcilable stakes of memory and return. The spectral, unconscious optics that emerge in her composite images blend the spaces of the individual journey of return with a larger global awareness of contested space and competing geopolitical interests.[25] Hers is an enlarged map that incorporates the losses of the Holocaust into a broader intertwined psychology and geography of irrecoverable loss. The illegibility of her works, moreover, and the multiple generations they have undergone in the process of copying and reproduction, signal the loss of materiality that the objects and images have suffered in the multiple expulsions they survived in faded and sometimes unrecognizable form.

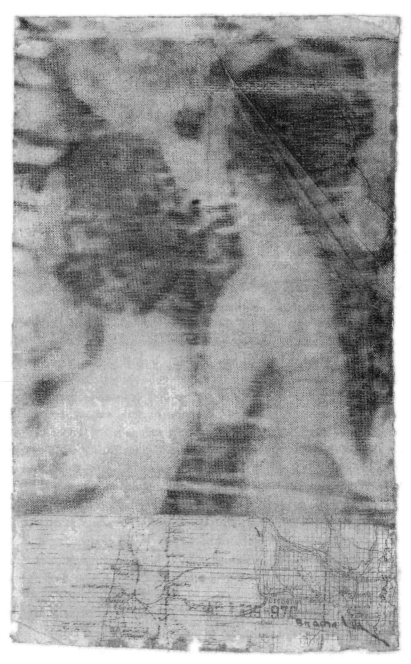

6.5 Bracha Lichtenberg-Ettinger, *Eurydice*, no. 2

18. Saidiya Hartman, *Lose Your Mother.*
19. Brett, *Too Many Men*, 526.
20. Ibid., 5.
21. Joseph Roach, *Cities of the Dead.*
22. The phantom sibling, or "memorial candle," is a ubiquitous and determining figure in narratives of massive trauma and loss. See, especially, Richieu in Art Spiegelman's 1986 *Maus*, but also, more recently, the figure of Simon in Phillippe Grimbert's novel and Claude Miller's 2008 film *A Secret.* On the notion of "memorial candle" see Dina Wardi, *Memorial Candle.* See also Hirsch, *The Generation of Postmemory.*
23. See especially Griselda Pollock, *Encounters in the Virtual Feminist Museum*; and Bracha Lichtenberg-Ettinger, "Matrix: Carnets 1985–1989" (artist book, 1992); *The Matrixial Gaze*; *Artworking, 1985–1999*; *The Eurydice Series*; Bracha Lichtenberg-Ettinger, with Emmanuel Levinas, *Que Dirait Eurydice? What Would Eurydice Say?*
24. Pollock, *Encounters*, 175.
25. On unconscious optics see Walter Benjamin, "The Work of Art in the Age of Mechanical Reproduction"; and Marianne Hirsch, *Family Frames.*
26. See Dalia Karpel ("With Thanks to Ghassan Kanafani") for a full account. Michael's novel has not been translated into English.
27. See Rebecca Harrison, "Review of 'The Return to Haifa.'"

BIBLIOGRAPHY

Assmann, Aleida. *Der lange Schatten der Vergangenheit: Erinnerungskultur und Geschichtspolitik.* Munich: C. H. Beck, 2006.

Benjamin, Walter. "The Work of Art in the Age of Mechanical Reproduction." In *Illuminations*, 217–52. Trans. Harry Zohn. New York: Schocken Books, 1968.

Brett, Lily. *Too Many Men.* New York: HarperCollins Perennial, 2002.

Connerton, Paul. *How Societies Remember.* Cambridge: Cambridge University Press, 1989.

Harrison, Rebecca. "Review of 'The Return to Haifa.'" *Ha'aretz*, 15 April 2008.

Hartman, Saidiya. *Lose Your Mother: A Journey Along the Atlantic Slave Route.* New York: Farrar, Straus and Giroux, 2007.

Hirsch, Marianne. *Family Frames: Photography, Narrative and Postmemory.* Cambridge: Harvard University Press, 1997.

——. *The Generation of Postmemory: Writing and Visual Culture After the Holocaust.* New York: Columbia University Press, 2012.

——, and Leo Spitzer. *Ghosts of Home: The Afterlife of Czernowitz in Jewish Memory.* Berkeley: University of California Press, 2010.

——, and Nancy K. Miller, eds. *Rites of Return: Diaspora Poetics and the Politics of Memory.* New York: Columbia University Press, 2011.

——, and Leo Spitzer. "The Tile Stove." *WSQ* 36, nos. 1 and 2 (Spring/Summer 2008), 141–50.

Huyssen, Andreas. *Present Pasts: Urban Palimpsests and the Politics of Memory*. Stanford: Stanford University Press, 2003.

Kanafani, Ghassan. *Palestine's Children: Return to Haifa and Other Stories*. Trans. Barbara Harlow. London: Heinemann, 2000.

Karpel, Dalia. "With Thanks to Ghassan Kanafani." *Ha'aretz*, 15 April 2005.

Lichtenberg-Ettinger, Bracha. *Artworking, 1985–1999*. Brussels: Palais des Beaux Arts, 1999.

——. *The Eurydice Series*. Ed. Catherine de Zegher and Brian Massumi. New York: The Drawing Center's Drawing Papers 24, 2001.

——. "Matrix: Carnets 1985–1989." Artist book, 1992.

——. *The Matrixial Gaze*. Leeds: Feminist Arts and Histories Network, 1995.

——, with Emmanuel Levinas. *Que Dirait Eurydice? What Would Eurydice Say? Emmanuel Levinas en/in Conversation avec/with Bracha Lichtenberg-Ettinger*. Amsterdam: Kabinet in the Stedeijk Musem, 1997.

Michael, Sami. *Doves in Trafalgar*, 2005.

Miller, Nancy K. "Emphasis Added: Plots and Plausibilities in Women's Fiction." *PMLA* 96, no. 1 (January 1981): 36–48.

Pollock, Griselda. *Encounters in the Virtual Feminist Museum: Time, Space and the Archive*. London: Routledge, 2007.

——. "Inscriptions in the Feminine." In Catherine de Zegher, ed., *Inside the Visible: An Elliptical Traverse of 20th Century Art, in, of, and from the Feminine*, 67–88. Cambridge: MIT Press, 1996.

——. "Rethinking the Artist in the Woman, the Woman in the Artists, and That Old Chestnut, the Gaze." In Carol Armstrong and Catherine de Zegher, eds., *Women Artists at the Millenium*, 35–84. Cambridge: MIT Press, 2006.

Roach, Joseph. *Cities of the Dead: Circum-Atlantic Performance*. New York: Columbia University Press, 1996.

Sa'di, Ahmad H., and Lila Abu-Lughod. *Nakba: Palestine, 1948, and the Claims of Memory*. New York: Columbia University Press, 2007.

Sebald, W. G. *Austerlitz*. Trans. Anthea Bell. New York: Modern Library, 2001.

Taylor, Diana. *The Archive and the Repertoire: Performing Cultural Memory in the Americas*. Durham: Duke University Press, 2003.

Wardi, Dina. *Memorial Candles: Children of the Holocaust*. New York: Routledge, 1992.

Young, James E. *The Texture of Memory: Holocaust Memorials and Meaning*. New Haven: Yale University Press, 1993.

7

NARRATIVES AND RIGHTS
ZLATA'S DIARY AND THE CIRCULATION
OF STORIES OF SUFFERING ETHNICITY

Sidonie Smith

A t this historical moment, human-rights activism remains the primary
global project for managing injustice and immiseration around the world,[1]
and life stories are at once ground and grist of rights work, rights instrumentali-
ties, and rights discourse. This conjunction of life narration, broadly defined,
and contemporary human-rights activism, is indeed, as Kay Schaffer and I argue
in *Human Rights and Narrated Lives: The Ethics of Recognition*, a productive and
problematic yoking of the decidedly intimate with the global.[2] Since the language of human rights is the contemporary lingua franca for addressing the
problem of suffering,[3] the attachment of personal storytelling to the discourse
and the institutions of human-rights activism enables survivors of and witnesses
to injury and harm to make their grievances public and to draw attention to
specific environments of suffering around the world. At the same time, this yoking of personal narrative and international rights politics affects the kinds of
life stories that can gain a global audience.

Take, for instance, the post–Cold War resurgence of ethnic nationalism,
with the attendant reorganization of politics in Eastern Europe, that set large
numbers of people in motion—into refugee camps, resettlement programs, and
diasporic communities in receiving nations such as the United States. Under violent assault, displaced, and haunted by traumatic memories, members of ethnic
communities turn to life storytelling to extend global recognition of the violence unleashed against people on the basis of their ethnic identification. Their
acts of narration emerge out of local contexts of rights violations; but to the
extent that local movements "go international," these witnesses participate

through personal storytelling in global processes that create a climate for the intelligibility, reception, and recognition of new stories about ethnicity under assault. Gillian Whitlock calls this breakthrough to public attention a "discursive threshold."[4]

Through their stories of ethnic suffering, witnesses expose the violence inflicted by those pursuing the project of ethnic nationalism as a goal of state formation. They also reveal the complexities and conundrums involved in telling stories of ethnic difference and grievance through frameworks and institutions founded on the concept of abstract universality. For many witnesses, the embeddedness of stories of ethnic suffering in the discourses, institutions, and practices of human rights provides the previously unheard and invisible a narrative framework, a context and occasion, an audience, and a subject position from which to make claims. And yet, in order to circulate their stories within the global circuits of human-rights activism and bring crises of violence and suffering to a larger public, witnesses give their stories over to journalists, publishers, publicity agents, marketers, and rights activists whose framings of personal narratives participate in the commodification of suffering, the reification of the universalized subject position of innocent victim, and the erasure of historical context and complexity through recourse to the feel-good opportunities of empathetic identification.

Personal storytelling in the context of rights activism is animated by and contributes to a paradox at the heart of human-rights discourse and practice: the uneasy enfolding of the universal in the ethnic particular. Elicited, framed, produced, circulated, and received within the contexts of human-rights activism, the life story of ethnic suffering at once ennobles an authentic (and sentimentalized) voice of suffering and depersonalizes that voice. Emerging from a local site of ethnically based struggle, the story enters Western-dominated global circuits, through which it may lose its local specificity. It reaches global audiences far from its point of origin, there to be interpreted and reproduced in unpredictable ways, some of which may elide difference as they universalize suffering and survival.

I cannot possibly do justice in this essay to the complexities of the conjunction of life storytelling and the contemporary geopolitics of human rights, as that conjunction captures and complicates transnational ethnic formations. What I can do is explore one narrative of besieged ethnicity, here Zlata Filipović's *Zlata's Diary*,[5] in order to elucidate some of the contradictory effects of the commodification of narratives of suffering ethnicity through rights activism that attaches abstract universality to ethnic difference under assault. I do so in order to assay what we might learn about the mobilization of personal stories of ethnic suffering in a globalized human-rights regime that serves as one of the central "managers" of ethnicity today.[6]

ZLATA'S DIARY IN CONTEXT

Reflecting on the dynamic relay between the ethnic, the national, and the diasporic, Ien Ang observes that "the rise of militant, separatist neonationalisms in Eastern Europe and elsewhere in the world signals an intensification of the appeal of ethnic absolutism and exclusionism which underpin the homeland myth, and which is based on the fantasy of a complete juncture of 'where you're from' and 'where you're at.'"[7] She goes on to note, as have Bruce Robbins and Elsa Stamatopoulou, the affective and political force of "the principle of nationalist universalism," or what she describes as "the fantasmatic vision of a new world order consisting of hundreds of self-contained, self-identical nations."[8] The struggle by ethnic nationalists to realize this new world order played out violently and traumatically in the former Yugoslavia in the 1990s. In the midst of the "new Europe," the people of the former Yugoslavia found themselves participants in a drama of salient ethnicities.

In the wake of President Tito's death and the loss of Soviet hegemony in Eastern Europe, the former Yugoslavia fractured into ethnic and (tenuously) multiethnic states. Croatia and Slovenia declared independence in 1991. In early March 1992, the government of Bosnia-Herzegovina held a referendum on independence from the Yugoslav federation (dominated by Serbia), which was boycotted by the Bosnian Serbs. The Bosnian parliament declared independence on 5 April 1992. Before official recognition of the decision by the European Community on 6 April, however, wars for ethnic dominance and hegemony erupted in Bosnia as secessionist Serb paramilitaries launched their campaign to gain control of the new country for annexation to the Republic of Serbia. Immediately, the city of Sarajevo came under siege, one that would last until a cease-fire went into effect in late 1995. Inhabitants of the multiethnic city of Sarajevo, which had gained world attention when it hosted the international athletes and spectators of the 1984 Olympic Games, found themselves trapped inside the blockaded city, forced to organize their everyday lives so as to evade sniper bullets and mortar attacks from surrounding mountains and to find scarce food and medical supplies. Bosnian Serb paramilitaries, supported by Slobodan Milošević in the Republic of Serbia, pursued a policy of genocide ("ethnic cleansing") for which they are now being held accountable in the international war crimes trials at The Hague. By early 1994 the United Nations reported that some ten thousand people had lost their lives or gone missing, among them fifteen hundred children; another fifty-six thousand had been wounded, including fifteen thousand children.[9] Eventually, the international community intervened, taking action against the besieging Serbs. As the Serbian paramilitaries lost ground, peace negotiations gained momentum. In October 1995, the United Nations brokered a cease-fire in Bosnia; in December the Dayton Accords were signed, establishing the blueprint for postwar stability, which would involve two autonomous governmental entities, the Muslim-Croat political entity called the Federation of Bosnia and Herzegovina and the Serbian government of the Republika Srpska. In late February 1996 the Bosnian government declared the siege of Sarajevo officially over.

Throughout the four-year siege, journalists assigned to Sarajevo reported on the realities of life lived under siege, the deaths of noncombatants, and the devastation of the city and its infrastructure. Writers and activists within Sarajevo also brought their stories of ethnic cleansing and ethnicity under assault to the wider world. Slatko Dizdarevic's *Sarajevo: A War Journal* (1993)[10] chronicled the early years of the siege. Then in late 1993 another personal story reached an international audience, this one the diary of a young girl.

From September 1991 to October 1993, the young Zlata Filipović kept a diary in which she recorded her everyday life in an increasingly besieged Sarajevo. Through her diaristic record of that everyday life, the young Bosnian-Croat described, and sometimes reflected upon, the disintegration of a cosmopolitan way of life and the gradual disruption and degradation of middle-class familiality through a war of ethnic nationalisms and the genocidal assault by Bosnian Serb paramilitaries. In the summer of 1993, Zlata shared her diary with her teacher, who subsequently found a publisher for it in Sarajevo. Through the sponsorship of the International Centre for Peace, the diary was originally published in Croat by UNICEF. With the recognition of the diary in Bosnia, Zlata became a "celebrity," labeled "the young Anne Frank" of Sarajevo. Recognizing the affective appeal of a story of the siege refracted through the eyes of the young girl, international journalists covering the war began reporting on Zlata and her story.

Several months after its publication in Sarajevo, a French photographer took a copy of Zlata's diary to Paris, where Le Robert Laffont-Fixot made a successful bid to become its French publisher. Le Robert Laffont-Fixot also provided the money and means to fly Zlata and her family from Sarajevo to Paris just before

Christmas 1993. In early 1994 the French translation of the diary appeared as *Journal de Zlata*. In this instance, life writing functioned as a means of life saving. The material diary became a commodity through which a life's sheer survival and betterment could be exchanged. Zlata's diary writing gained her and her family escape from snipers and the bombs of the siege and enabled her to start a new life in Paris (and subsequently Ireland).

From Paris the diary traveled to New York City, where it was auctioned off in a sale conducted by its French publisher. With a bid of $560,000, Viking Penguin (a subsidiary of Penguin Books) won the rights to publish *Zlata's Diary: A Child's Life in Sarajevo* and did so in March 1994. After publication in the United States, the diary began to reach an ever-widening readership. Irene Webb of International Creative Management subsequently bought the rights to represent the book in any movie deal. As reported in the *New York Times* on 19 January 1994, Webb announced: "It's like the 'Diary of Anne Frank,' but with a happy ending."[11] Upon publication, the English-language version of the diary circulated broadly within the United States, becoming "an extraordinary national bestseller," according to the book cover. Eventually the diary moved into the social studies curricula in the nation's public schools. As one Web site announces: "Zlata's diary brings Sarajevo home as no news report ever could"[12] (The Unsung Heroes of Dialogue).

In the years after the diary's publication, Zlata became a "spokeschild" for the conditions of ethnic genocide and displacement in the former Yugoslavia, appearing through the auspices of the United Nations as an ambassador speaking on behalf of the children of Bosnia. International attention brought increased interest in conditions in the former Yugoslavia and, after the cessation of fighting, in the rights of the child internationally. In 1995 Zlata appeared as a special guest at the 1995 Children's World Peace Festival in San Francisco. The attention garnered by the diary and its circulation within the United States and Europe produced an aura-effect around Zlata herself, elevating her and legitimating her as a "universal" voice of the child suffering from human-rights abuses. Since those years, Zlata has continued as an activist on behalf of the human rights of children, helping to launch UNICEF reports on the impact of armed conflict on children. Her diary continues to be highlighted as suggested reading on Web sites mounted by activists working on behalf of the UN Convention on the Rights of the Child. During the summer of 2004, a stage version of *Zlata's Diary*, produced by Communicado Productions, toured Scotland. Though no longer a child, Zlata continues to speak on behalf of besieged childhood from her home in Dublin.

REFLECTIONS ON ZLATA'S STORY OF SUFFERING ETHNICITY

Interethnic Appeals and the Production of Collective Memory

Through the publication of the diary in the West, "Zlata" becomes a marketable archetype of the suffering victim of ethnic nationalism in extremis. The publication of her story of lost childhood, of innocence under assault, is meant to lend immediacy to calls for intervention on the part of the international community in the ethnic war in Bosnia and the organized acts of genocide carried out in service to nationalist myths and the nationalist "fantasy of a utopic space to be occupied by all those who suffered 'the same' violence at the hands of the enemy."[13] And yet the case of Zlata's diary suggests how interethnic the appeal to ethnicity under assault becomes.

Within Zlata's diary and within its zones of circulation and reception, Jewish ethnicity comes to underwrite the aura of suffering of a largely unmarked Croatian ethnicity. Here is an instance in which one ethnicity gets attached to another ethnicity globalized in world memory through a particular mode of life writing, the child's diary. Zlata herself invokes the comparison to Anne Frank early in her *Diary*. Like Anne Frank, she chooses a name for her diary. Writing on Monday, 30 March 1992, she opens her entry with, "Hey, Diary! You know what I think? Since Anne Frank called her diary Kitty, maybe I could give you a name too."[14] In all subsequent entries Zlata addresses the diary as Mimmy, projecting an affectionate and interested interlocutor and keeper of secrets. References to Anne Frank and her diary are not only intertextual, they are also extratextual. Already in 1993 in Sarajevo, Zlata, at thirteen, was being labeled "the young Anne Frank."[15] The identification of Zlata's story with the story of Anne Frank, its modeling upon the earlier text, its adoption of the earlier diarist's mode of address to an interlocutor—all suggest the way in which the "authenticity" of this contemporary girl, this "Zlata," derives from the earlier editing and marketing of "Anne Frank" as a figure of universalized innocence and heroic suffering whose celebrity can be borrowed in making claims about the struggle against racial violence and ethnic cleansing.

In Zlata's self-positioning in her diary as a modern-day Anne Frank and in the marketing of "Zlata" as a new "Anne Frank," both narrator and marketer assume the global resonance of the iconic figure of Anne Frank; they assume that "Anne Frank" will be collectively remembered as having been tragically lost in the Holocaust. The affective appeal of the Sarajevan girl's story of lost childhood becomes intelligible to a broad educated readership through the global aftereffects of collective world memory of another "ethnic" girl's narrative of lost girlhood and lost life. The haunting remains of "Anne Frank," and the aura of the Holocaust as a paradigmatic event of twentieth-century genocide, attaches itself to this "child's" narrative as Zlata and her publishers attach her story to that of Anne Frank, who has through her widely read *Diary*, as the Web site for the Anne Frank Foundation puts it, "become a world-wide symbol representing all victims of racism, anti-Semitism and fascism. She stands for victims who lived at the same time as she did just as much as for the victims of today. The foremost message contained in her *Diary* sets out to combat all forms of racism and anti-Semitism" (Anne Frank Foundation).[16] In this instance, Jewish ethnicity functions as an ethnicity of reference in the globalization of human-rights activism.

The forces of globalization, Clifford Bob notes, offer victims and activists responding to ethnic violence "symbols of oppression and repertoires of contention" through which to organize and project their local grievances in an international arena.[17] This intertextual aspect of the production and circulation of *Zlata's Diary* within the rights context points to ethnic remembering and storytelling as an historical effect of transethnic comparison, an interethnic energy distributed across unevenly remembered events in world memory. Brent Edwards argues that "the level of the international is accessed unevenly by subjects with different historical relations to the nation."[18] I would adapt his argument to make the point that the level of the transnational is accessed unevenly by ethnic subjects with different historical relations to the global circuits of world memory.

The Depoliticization of a Globalized Ethnic Suffering

In her critique of the sentimentalization of suffering, Karyn Ball has called for the comparative study of traumatic histories, in order, she writes, "to forge links among traumatic histories that would raise Americans' historical consciousness and promote their sense of civic responsibility."[19] Ball's is a call for comparative studies of histories of suffering, necessary to complicating any one model of traumatic remembering, any one paradigm for understanding witness testimony, and any singular model of possibilities for recovery and recognition. In one sense we might read *Zlata's Diary* as pursuing, in its yoking of "Zlata" and "Anne Frank," what Ball describes as a "strategy of comparison in order to forge links among traumatic histories" Here is a strategy of comparison in action. And yet this strategy of comparison from the ground up, as it were, and through the perspective of an adolescent immersed in a globalized popular culture, may not so much illuminate the incommensurable differences and the specificities of ethnic histories as effect a flattening of historical specificity through an appeal to empathetic, depoliticized sentimentality.

As a text commanding response and responsible action, *Zlata's Diary* is represented and marketed in ways that sentimentalize the suffering Bosnian-Croat subject by lifting that subject outside history and politics. The commodification of stories of ethnic suffering obscures the complex politics of historical events, stylizes the story to suit an educated international audience familiar with narratives of individual triumph over adversity, evokes emotive responses trained on the feel-good qualities of successful resolution, and often universalizes the story of suffering so as to erase incommensurable differences operative amidst the horror of violence. The commodification of the young girl's diary gives us a version of the story of "Anne Frank"—but with a happy ending.

Yet there is more to the relationship established between the contemporary Zlata and the 1940s Anne Frank. The forces of commodification have framed the earlier diary as well. In successive decades since its initial publication, *The Diary of Anne Frank* has been edited and interpreted, reedited and reinterpreted, marketed and circulated, to give some of its audiences an "Americanized" "Anne Frank" situated not in a determinative ethnicity but situated as an adolescent subject inspiring hope and promise "for everyone." As an early reviewer of the stage version of the Diary wrote in 1955, "Anne Frank is a Little Orphan Annie brought into vibrant life."[20] Alvin H. Rosenfeld suggests that the early version of the diary, and the 1955 stage play based on the diary as adapted by Frances Goodrich and Albert Hackett, present "an image of Anne Frank that would be widely acceptable to large numbers of people in the postwar period . . . one characterized by such irrepressible hope and tenacious optimism as to overcome any final sense of a cruel end."[21] He further elaborates how the play and its reviews erase the haunting marks of ethnic difference, eliding references to the Jewishness of the Frank family and playing up the figure of the "universal" teenager struggling in adolescence and hopeful about the future.

The Jewish particularities of the Anne Frank who lived in the attic and died in Bergen Belsen are suppressed in order to broadcast a story of universal inspiration. Made into a story that "speaks" to "everyone" about what Hanno Loewy describes as "the personalized world of family experience,"[22] the diary of Anne Frank becomes a story that can no longer speak of ethnic difference. The iconic "Anne Frank" becomes an abstract universal "detached from her own vivid sense of herself as a Jew."[23] Rosenfeld defines Anne Frank as a "contemporary cultural icon" whose name is so well known that "to the world at large" the 1.5 million children who perished in the Holocaust "all bear one name—that of Anne Frank."[24] "Anne Frank" has become the child that died in the genocidal Holocaust.

Decades later, the production, circulation, and reception of Zlata's story of ethnicity under assault as "the deepest truth about the Bosnia situation" had the effect of "leech[ing]," according to David Rieff, "the Bosnian tragedy of its complexity."[25] If the category of the ethnic, and the global visibility and saliency of particular ethnic identifications, are historical effects of a modernity founded on the articulation of universal categories of abstract equality,[26] then the trackings of ethnicities enfolded in one another at once create a superfluity of the particularities of difference and cancel differences through the abstract equality (or universalism) of those who share suffering. The figure of the child commodified in the global flows of rights activism and its management of ethnicity becomes the sentimental public face of ethnic trauma and the violence of ethnic nationalism, the essentialized figure of the community's "victim" and its victimization. To put it another way, "Zlata," with her invocation of "Anne Frank," becomes, for some readers, the universalized figure of ethnicity's besieged child.

The Remains of Ethnic Suffering

And yet, for other readers, and particularly members of communities experiencing contemporary displacement, ethnicity can function as a trace of continuity across rupture. In this context, stories of ethnic suffering can offer occasions for constituting the remembered past as a resource for understanding identities in the social present.[27] When Zlata's Diary enters into circuits of consumption in the United States and Western Europe, the narrative begins to circulate in venues where it can be invoked as a marker of Croatian ethnicity under assault, or a lost Bosnian cosmopolitanism, thereby sustaining nationalist narratives of suffering and loss so often central to the imagination of the ethnic as a site of sentimental attachment. Because reading narratives of suffering and loss is not only "a profoundly personal act, belonging to a psychological sphere, but . . . also the effect of inhabiting various cultural spaces,"[28] published narratives such as this diary produce an archive of memories of "ethnicity." The story might thereby set in motion new releases of affective energies;[29] and those energies can be put to use in the social struggles over competing rememberings of "Bosnia" and its wars of ethnic nationalisms. This story can become a part of a cultural inventory, the reservoir of collective memory upon which ethnic nationalism is both founded and sustained.

For some, then, *Zlata's Diary* participates in the production and circulation of new collective memories (for members of the diasporan Croatian community in the United States, for instance), offering a future site of melancholy, what David Eng and David Kazanjian define as the "psychic and material practices of loss and its remains."[30] It puts in play residual glimpses of the past as remembered tradition of interethnic community or ethnic grievance. Contributing to "a contemporary landscape of memory"[31] through which future subjects may negotiate their ethnic attachments and pasts, the personal diary may underwrite future historical grievances. This narrative of loss told through the voice of the "innocent" child becomes a site of melancholy that "creates a realm of traces open to signification, a hermeneutic domain of what remains of loss."[32] The child becomes the sentimental public figure of ethnic trauma.

The Universalized Innocence Effect

In the context of human-rights campaigns, life narrators are expected to take up the subject position of "innocent" victims; and they are expected to be able to occupy that position with moral authority. Yet, some survivors of human-rights abuses are more easily equated with the subject position of victim than others. The child is, as Tony Hughes D'aeth suggests in his discussion of the film *Rabbit-Proof Fence* (2002), easily the most accessible and readily believable of victim identities.[33] The marketing of Zlata as a victim/commentator on suffering ethnicity situates the young girl in the subject position of unassailable innocent. This marketing of sentimental suffering, through a child's-eye narrative viewpoint and the trope of childhood lost, reinforces the differentiated identities of ethnic victim and ethnic perpetrator, and secures rather than renders complex the moral alignment of innocence and victimization. Begging the question of innocence in childhood, we might say that in human-rights discourse and campaigns, "the child" is given to speak for the better part of "ourselves," the better part of human nature, the better part of our community. In this way, the child is positioned as the voice of knowledge, as one speaking wisdom, an effect that Rieff, in his critique of *Zlata's Diary*, assails and distrusts.[34]

The "innocence" effect is produced through Zlata's self-conscious invocation of the trope of lost childhood and her shifting terms of reference. Within her diary, Zlata is self-conscious about the importance of her narrative and its possible importance to others. She even writes about becoming a "personality" after the initial publication of the diary. And once interest is expressed in her diary, she begins to reflect on the situation in Sarajevo in rather poignant ways. Zlata's self-consciousness about her celebrity and her recognition of her role as representative child of Sarajevo, and about the tragedy of "lost childhood" (a discourse that comes from the journalists and advocates who take up her story), undoes the truth effect of "innocent child" and the "child's-eye view" otherwise produced through the diary. Already, within the production of the diary, the politics of commodified sentimentality are evident.

The innocence effect is also reinforced through the packaging of *Zlata's Diary* and the paratextual use of photographs that visualize the young girl's story as a sentimentalized drama of lost childhood. One photograph in particular captures "innocence" and the "production of innocence" at the same time. This is the photo of Zlata in bed, framed by the caption "Zlata, who loves books, reads by candlelight." To get the picture for mass distribution through global media, the candlelight has to be photographed; photographed, it is overwhelmed and rendered inauthentic by the light from the flash of the bulb. Through such visuals, the authenticity of sentimental childhood is at once produced and exposed as artificial, as the reviewer for *Newsweek* noted.[35] My point here is not that the diary is "inauthentic" or "suspect" as witness testimony. It is, rather, that the commercialization of the diary and its capture in what Lauren Berlant has described as "sentimental politics," here a politics of the ethical (acts of soliciting response and responsibility across social divides), obscures the difficult politics of histories of difference and violence.[36]

Saving Whose Child

In her preface to the diary, journalist Janine Di Giovanni prepares the reader for this child witness and her story, thereby offering the text's first reading. Zlata, she writes: "kept a careful record of the chilling events—the deaths, the mutilations, the sufferings. When we read her diaries, we think of desperation, of confusion and of innocence lost, because a child should not be seeing, should not be living with this kind of horror. Her tragedy becomes our tragedy because we know what is happening in Sarajevo. And still, we do not act."[37] Through a reading praxis that foregrounds the figure of the innocent child and the trope of innocence lost, Di Giovanni orients a global middle-class readership to the "representative" story of all the suffering children of Sarajevo. In this introduction, as elsewhere in the era of humanitarianism and human rights, images of children and lost childhoods are invoked to shame individuals, communities, nations, and the imagined "international community" into action. Those images become invitations to rescue. In this call to affect and action, the reader is addressed as both passive bystander and surrogate parent, shamed and enjoined to respond as the parent of all children.

Critically, the diary presents a universal child of a normative middle-class family. And it is marketed to a broad middle-class readership educated about, familiar with, and prepared to respond to pleas for good parenting. The published version includes a cast of family characters and a photo album, with images of the wholesome, open-faced, smiling Zlata, a figure of the innocent child tugging on the sympathies of the reader. The home imaged is a middle-class one, the occasions of the photos birthday parties and family outings. The photo album appeals to Western readers—both adults and children, presenting a home and a family the educated reader can imagine inhabiting. Additionally, throughout the diary, Zlata's references to global popular culture resonate with the lives, desires, and habits of young people in Western Europe and the United States. The constant citation of a global popular culture (a popular culture whose primary, though by no means whole, point of reference is the United States) situates the subject of the diary in a nondifferentiated space of consumer adolescence and global youth culture. In this, Zlata is "representative" of a commodified and "universal" middle-class adolescent subject knowledgeable about and attentive to the products, icons, celebrities, and self-descriptions of the global marketplace.

As she interweaves comments on the common references of global youth culture and the trope of childhood lost, Zlata assumes the subject position of the universal middle-class child anxious about childhood itself.[38] Lisa Makman has observed that children themselves have now become the crusading upholders of the rights of the child to a childhood perceived by increasing numbers of people in industrialized democracies as under assault. Makman tracks recent UN discourse about "the world's children" and attributes this focus on childhood under assault to cultural anxieties, circulating in the mainstream media in the United States, about the "ero[sion]" of a "universal" innocent childhood caused by the influences of new technologies and global media.[39] Through the commodification of stories of ethnic suffering and the sentimentalized "channels of affective identification and empathy,"[40] ethnicity's besieged child is becoming the universally besieged child of a universally besieged childhood. In effect, the diary, the paratextual apparatus surrounding it, and its marketing register, not the particularity of the ethnic subject under duress, but rather the abstract universality of the child of human rights; and one effect of the appeal to "the child" in need of saving is that the child is everybody's child and thus nobody's specific child.

CONCLUSION

As a marker of identity and difference, ethnicity is an effect of modernity rather than a residue from the past prior to modernity. Ethnicity, Jack David Eller suggests, is "a radical appropriation and application of otherness to the practical domain."[41] Thus, modernity involves what Rey Chow describes as "the systematic codification and management of ethnicity."[42] The contemporary human-rights arena is a primary site for this project of codification and management. In human-rights campaigns targeted at violation of ethnic rights in the midst of ethnic nationalism in extremis, ethnicity has to be "managed" as immobile difference through a modernist fiction of a totalizing ethnicity (a definitive inside to a collectivity) under assault from an outside.[43] In such arenas, human-rights discourse and campaigns are responses to, and in turn engage in, the production of salient ethnicities and ethnicities of reference.

The case study of *Zlata's Diary* and the problematics of ethnic suffering exposes the "logical contradictions" and "epistemic paradoxes"[44] enfolded within and enfolding the production, circulation, and reception of personal narratives in the regime of human rights. Abstract universality and ethnic difference are both "mythographic reductions" at once underwriting, energizing, and reconfiguring human-rights discourse and activism. Through the pathways and byways of global circuits localized and local circuits globalized, the tensions binding abstract universality and ethnic difference release energies that reconnect, diverge, and converge around the international community's struggle with injustice and suffering. I have used *Zlata's Diary* to reflect on the circuits of ethnicity as sentimental politics. And so let me conclude with some observations about the conjunction of human rights and narrations of suffering ethnicity.

Narratives enlisted in and attached to human-rights campaigns participate in the articulation of a history of suffering and loss attached to ethnic identity and the articulation of communal fictions of ethnicity (imaginings and grievances). Narratives of suffering and loss bind communities sharing some common "ethnic" past (or language, culture, defining events), across their local, national, and diasporic differences at the same time that they appeal to others who do not share that ethnic marker. They provide historical information, intergenerational communication, rallying cries, sites of healing. They offer a means to claim rights and demand redress and also to claim a shared past and shared tradition. They ignite an affective charge attached to identity under assault, project a figure of the victim for political mobilization, and serve as a means of shaming a nation and the international community into acknowledging and redressing claims. Because they are so critical to contemporary human-rights activism, stories such as that presented in *Zlata's Diary* become cultural capital, for individuals and for ethnic communities. Sometimes the publication and circulation of a specific narrative becomes a "focusing event"[45] that galvanizes international attention and action, as was the case with Rigoberta Menchú's *I, Rigoberta Menchú: An Indian Woman in Guatemala* (1984), which gained recognition for the situation of Guatemala's indigenous community in its struggle against a repressive state.[46]

Emanating from local settings that are inflected by and inflect the global, life stories are taken up in a host of different sites and networks—formal and informal, material and symbolic—where they undergo further transformations and remediations. In effect, narratives of suffering such as *Zlata's Diary* are produced, circulated, and received within an intricate, uneven, and overlapping set of spheres: the local, the national, the regional, the global. They also travel within overlapping, uneven, and intersecting zones of ethnic identification and affiliation: the diasporic, the transethnic; the national ethnic, and the local ethnic—all heterogeneous zones of identification and historical tracings differently located, differently accessed. Moreover, such stories, as we have seen, unfold through and enfold overlapping, uneven, and contradictory appeals to ethnic singularity and abstract universality at once.

Finally, commodified narratives of suffering ethnicity enter a global field saturated with multiple modes of appeal and cues to interpretation. They reach for readers/viewers/the public, calling that public into definition—as a middle-class public of parents and children or as an ethnic public of dispersed Bosnian Croat refugees, for instance. As with all such appeals, suggests Thomas Keenan, "the public is the possibility of being a target and of being missed."[47] For publics are themselves moving targets. They are sometimes galvanized through affective attachment to personal stories, and are sometimes indifferent to appeals. And since publics are multiple, readings of ethnicity under assault are heterogeneous, sometimes contradictory, and often unmanageable.

NOTES

I am indebted to Kay Schaffer for her comments on certain aspects of the framing and marketing of *Zlata's Diary*. I am indebted to Laurie McNeill for conversation about Anne Frank's diary. I am also indebted to John Cords and Elspeth Healey for their research assistance.

1. Paul Farmer, *Pathologies of Power*, 49.
2. Kay Schaffer and Sidonie Smith, *Human Rights and Narrated Lives*. In our study, we look expansively at the multiple sites of personal storytelling attached to human rights campaigns: published life narratives, fact-finding in the field, handbooks and Web sites, nationally based human rights commissions, human rights commission reports, collections of testimonies, stories in the media, and the scattered everyday venues through which narratives circulate.
3. Michael Ignatieff, *Human Rights as Politics and Idolatry*, 7.
4. Gillian Whitlock, *The Intimate Empire*, 144.
5. Zlata Filipović, *Zlata's Diary*.
6. See Rey Chow on the management of ethnicity, "Introduction: On Chineseness as a Theoretical Problem," 11.
7. Ien Ang, *On Not Speaking Chinese*, 34.
8. Ang, *On Not Speaking Chinese*, 34; Bruce Robbins and Elsa Stamatopoulou, "Reflections on Culture and Cultural Rights," 425.
9. See the United Nations, *Study of the Battle and Siege of Sarajevo*.
10. Slatko Dizdarevic, *Sarajevo: A War Journal*.
11. Sarah Lyall, "Auction of a War Diary."
12. The Unsung Heroes of Dialogue.
13. Richard Wilson, "The Sizwe Will Not Go Away," 16. Wilson here cites the work of Glenn Bowman and his discussion of the ways in which "the narrating of past mass violations plays a constitutive role in the formation of all nationalisms" (quoted in Wilson, 16). Wilson does not give a reference for the Bowman paper.
14. Filipović, *Zlata's Diary*, 27.
15. Janine Di Giovanni, Introduction to *Zlata's Diary*, v.
16. Anne Frank Foundation.
17. Clifford Bob, "Globalization and the Social Construction of Human Rights Campaigns," 134.
18. Brent Edwards, *The Practice of Diaspora*, 7.
19. Karyn Ball, "Trauma and Its Institutional Destinies," 15.
20. *New York Daily News*, 6 October 1955, quoted in Alvin H. Rosenfeld, "Popularization and Memory."
21. Ibid., 251–52.
22. Hanno Loewy, "Saving the Child," 156.
23. Rosenfeld, "Popularization and Memory," 257.
24. Rosenfeld, "Popularization and Memory," 244, 243. As Alvin Rosenfeld makes clear, "Anne Frank" is remembered differently in different communities at different historical moments. After analyzing the Americanization of "Anne Frank," he goes on to explore the reception of the diary by Germans and by Jewish writers and intellectuals and concludes that "in both Germany and Israel one finds a common history marked by a common symbol but shaped by very different motives and yielding diverse interpretations of the past" (277).
25. David Rieff, "Youth and Consequences," 32.

26. See David Kazanjian, *The Colonizing Trick*, 4–27.

27. Jack David Eller, "Ethnicity, Culture, and 'the Past.'"

28. Jill Bennett and Roseanne Kennedy, *World Memory*, 7.

29. Félix Guattari, *The Three Ecologies*, 36.

30. David L. Eng and David Kazanjian, *Loss*, 5.

31. Bennett and Kennedy, *World Memory*, 8.

32. Eng and Kazanjian, *Loss*, 4.

33. Tony Hughes D'aeth, "Which Rabbit-Proof Fence?."

34. Rieff, "Youth and Consequences," 33. David Rieff also indicts the way she is made to speak as a commentator on behalf of Bosnian innocence. Comparing the versions of the diary published in Paris and in the United States and the interpolations added to the Viking Penguin edition, he notes the addition of references to political events and critiques of the leaders and their actions (33–34).

35. "Child of War: The Diary of Zlata Filipovich."

36. Lauren Berlant, "The Subject of True Feeling."

37. Di Giovanni, "Introduction," xii.

38. Stuart Hall has cautioned that cultural formations may work in contradictory ways. There is at once the force of homogenization and universalization across national and ethnic differences through appeals to global mass culture. There is also the incorporation and reflection back through global mass culture of the specific context of ethnic difference and its histories of suffering (Hall, "The Local and the Global"). I may be overstating the former case here.

39. Lisa Hermine Makman, "Child Crusaders," 289, 291.

40. Berlant, "The Subject of True Feeling," 53.

41. Eller, "Ethnicity, Culture, and 'the Past.'"

42. Chow, "Introduction," 11.

43. See Chow, "Introduction."

44. Djelal Kadir, "Introduction: America and Its Studies," 14.

45. Bob, "Globalization and the Social Construction of Human Rights Campaigns," 136, citing John Kingdon, *Agendas, Alternatives, and Public Policies*, 99–100.

46. For a discussion of the publication and reception of Menchú's narrative, see Schaffer and Smith (*Human Rights and Narrated Lives*) and the essays in Arias (*The Rigoberta Menchú Controversy*).

47. Keenan, "Publicity and Indifference," 108.

BIBLIOGRAPHY

Ang, Ien. *On Not Speaking Chinese: Living Between Asia and the West*. London: Routledge, 2001.

Anne Frank Foundation. *Responsibilities of the Anne Frank-Fonds*. http://www.annefrank.ch (accessed 5 December 2011).

Arias, Arturo, ed. *The Rigoberta Menchú Controversy*. With a response by David Stoll. Minneapolis: University of Minnesota Press, 2001.

Ball, Karyn. "Trauma and Its Institutional Destinies." *Cultural Critique* 46 (Fall 2000): 1–44.

Bennett, Jill, and Roseanne Kennedy. *World Memory: Personal Trajectories in Global Time*. London: Palgrave, 2003.

Berlant, Lauren. "The Subject of True Feeling: Pain, Privacy, and Politics." In Austin Sarat and Thomas R. Kearns, eds., *Cultural Pluralism, Identity Politics, and the Law*, 49–84. Ann Arbor: University of Michigan Press, 1999.

Bob, Clifford. "Globalization and the Social Construction of Human Rights Campaigns." In Alison Brysk, ed., *Globalization and Human Rights*, 133–47. Berkeley: University of California Press, 2002.

"Child of War: The Diary of Zlata Filipović." *Newsweek*, 28 February 1994, 24–27.

Chow, Rey. "Introduction: On Chineseness as a Theoretical Problem." *Boundary 2* 25, no. 3 (1998): 1–24.

Di Giovanni, Janine. "Introduction." In Zlata Filipović, *Zlata's Diary: A Child's Life in Sarajevo*. New York: Penguin, 1994.

Dizdarevic, Slatko. *Sarajevo: A War Journal*. Trans. Anselm Hollo. New York: Fromm International, 1993.

Edwards, Brent. *The Practice of Diaspora: Literature, Translation, and the Rise of Black Internationalism*. Cambridge: Harvard University Press, 2003.

Eller, Jack David. "Ethnicity, Culture, and 'the Past.'" *Michigan Quarterly Review* 36, no. 4 (1997): 552–601.

Eng, David L., and David Kazanjian. *Loss: The Politics of Mourning*. Berkeley: University of California Press, 2003.

Farmer, Paul. *Pathologies of Power: Health, Human Rights, and the New War on the Poor*. Berkeley: University of California Press, 2003.

Filipović, Zlata. *Zlata's Diary: A Child's Life in Sarajevo*. New York: Penguin, 1994.

Guattari, Félix. *The Three Ecologies*. Trans. Ian Pindar and Paul Sutton. London: Athlone, 2000.

Hall, Stuart. "The Local and the Global: Globalization and Ethnicity." In Anthony D. King, ed., *Culture, Globalization, and the World-System: Contemporary Conditions for the Representation of Identity*, 19–40. Minneapolis: University of Minnesota Press, 1997.

Hughes D'aeth, Tony. "Which Rabbit-Proof Fence? Empathy, Assimilation, Hollywood." *Australian Humanities Review*, September 2002. http://www.lib.latrobe.edu.au/AHR/archive/Issue-September-2002/hughesdaeth.htm (accessed 5 December 2011).

Ignatieff, Michael. *Human Rights as Politics and Idolatry*. Princeton: Princeton University Press, 2001.

Kadir, Djelal. "Introduction: America and Its Studies." *PMLA* 118, no. 1 (2003): 9–24.

Kazanjian, David. *The Colonizing Trick: National Culture and Imperial Citizenship in Early America*. Minneapolis: University of Minnesota Press, 2003.

Keenan, Thomas. "Publicity and Indifference (Sarajevo on Television)." *PMLA* 117, no. 1 (2002): 104–16.

Kingdon, John. *Agendas, Alternatives, and Public Policies.* 2nd ed. New York: HarperCollins, 1984.

Loewy, Hanno. "Saving the Child: The 'Universalisation' of Anne Frank." Trans. Russell West. In Rachael Langford and Russell West, eds., *Marginal Voices, Marginal Forms: Diaries in European Literature and History,* 156–74. Amsterdam: Rodopi, 1999.

Lyall, Sarah. "Auction of a War Diary." *New York Times,* 19 January 1994, C20.

Makman, Lisa Hermine. "Child Crusaders: The Literature of Global Childhood." *The Lion and the Unicorn* 26 (2002): 287–304.

Menchú, Rigoberta. *I, Rigoberta Menchú: An Indian Woman in Guatemala.* London: Verso, 1984.

Rieff, David. "Youth and Consequences: Zlata's Diary: A Child's Life in Sarajevo." *New Republic,* 28 March 1994, 31–35.

Robbins, Bruce, and Eisa Stamatopoulou. "Reflections on Culture and Cultural Rights." *South Atlantic Quarterly* 103, no. 2/3 (2004): 419–34.

Rosenfeld, Alvin H. "Popularization and Memory: The Case of Anne Frank." In Peter Hayes, ed., *Lessons and Legacies: The Meaning of the Holocaust in a Changing World,* 243–78. Evanston: Northwestern University Press, 1991.

Schaffer, Kay, and Sidonie Smith. *Human Rights and Narrated Lives: The Ethics of Recognition.* New York: Palgrave Macmillan, 2004.

United Nations. Study of the Battle and Siege of Sarajevo: Final Report of the United Nations Commission of Experts. 1994. http://www.ess.uwe.ac.Uk/comexpert/ANX/VI-01.htm#I.C. (accessed 5 December 201).

The Unsung Heroes of Dialogue. http://www.un.org/Dialogue/heroes.htm (accessed 5 December 2011).

Whitlock, Gillian. *The Intimate Empire: Reading Women's Autobiography.* London: Cassell, 2000.

Wilson, Richard A. "The Sizwe Will Not Go Away: The Truth and Reconciliation Commission, Human Rights, and Nation-Building in South Africa." *African Studies* 55, no. 2 (1996): 1–20.

8

LETTER FROM ARGENTINA

Nancy K. Miller

For a long time I had wondered about a letter in a thin, pale blue airmail envelope, frayed around the edges, and missing a stamp. The letter was addressed jointly to my father's brother, Mr. S. Kipnis, and to my grandmother, Sadie, in North Bergen, New Jersey. My grandmother's name spelled "Sade," was written off to the side, twice, one above the other on the envelope; the family name had also been added below, a second time, spelled Kipnes. On the envelope, the town's name was Hispanisized as Norte-Bergen, as were the rest of the address lines: New-York, America Norte. Amazingly, the letter was not returned to sender.

But who was the sender? The return address only compounded the mystery. It began with the notation Asociacion Israelita, which was followed on the next line by a family surname I didn't recognize. The rest of the return address read Entre-Rios 1471, Rosario (St. Fe). Why had Sadie, and then my father, saved this letter from Argentina (I found the letter among his papers after his death)? What was so important about it? I had no idea. But in a gesture that I have now come to acknowledge as a peculiar pattern of deferral in the detective work that uncovering the history of my father's side of the family has involved, when I saw that the letter was written in Yiddish, I set it aside for another day. I was curious, yes, but somehow not curious enough to engage a translator for the four pages written in a language of which I could not decipher a single word.

The obstacle wasn't the language per se. It was more the slightly perverse effect produced by a kind of unconscious filter. Despite the fact that I felt hun-

8.1 The envelope from Argentina

gry for whatever crumbs were left along the genealogical trail I was following, I would only pick up certain clues at any given moment. Then I would find myself, sometimes years later, as I am now, perplexed by my earlier, almost perverse blindness. On the one hand, I was grasping for any information that would tell me something about my father's side of the family; on the other, I would often miss what was staring me in the face, if the evidence didn't bear his name, as it were.

What fascinated me in the direct aftermath of my father's death was a family photograph, the group portrait I was later to begin to decipher with my cousin Sarah. At the time, I could identify only my Kipnis grandparents and my uncle (about age nine), but none of the five other people pictured. I turned then to my father's cousin, the only living relative I had contact with, one of two sisters who lived in Montreal, and whom I had met not long before his death. His mother and their mother were sisters. My father was especially close to Gert, the younger of the sisters, whom I had once met when she visited my father in New York in the 1980s. I liked her, and was touched by her affection for my father. Not long after he died, I sent her a copy of the photograph asking if she could help me figure out who was in the portrait. When she replied, she said that she didn't know the people in the picture—except, of course, for my grandmother, who was her own mother's sister. She told me something about them—the sisters— whatever she knew. Nothing she said registered; I didn't care about my grand- mother's side of the story then, only the genealogy of the Kipnis line. I filed the letter without pursuing the matter any further. You would think that I had learned no lessons about patriarchy and the name of the father in the course of my long feminist education.

Almost twenty years after my father's death, I finally had the letter trans- lated. It suddenly seemed stupid to keep documents in my possession that I couldn't read. And despite my earlier indifference, now that I was fully engaged in my quest, I was open to everything, though it's true I was also hoping that when translated into English the letter would reveal family secrets, about what I could not precisely say. I longed for something if not shocking, at least infor- mative that would produce a more detailed narrative of family history beyond the formulaic one I had inherited: "They left Russia because of the pogroms." Maybe the letter would flesh out my portfolio of silenced stories.

On the last page of the letter, in a convoluted turn of Yiddish *politesse*, the writer identifies herself as my grandmother's sister: "Please extend the most friendly and dearest of greetings to your loving and dear parents, they should live a long life, and extend to your loving mother that she has a sister that was given as a present to her because no one expected her to survive this ordeal. My friendliest greetings to your loving brother"—and she refers to my father, her nephew, by his Yiddish name.

The letter, two double-sided, yellowed pages covered in neat blue handwrit- ing, is a blow by-blow account of an operation. "From my belly, they took out such troubles and they left it in the hospital, so they could study it. Ten doctors were present at the operation, including students." My great-aunt's letter is one long narrative of pain relieved; a celebration of survival, thanks to God's inter- vention. Survival but not good health, not so fast: "And don't think that I can eat everything. They also need to warm my liver with electricity. I need lots of warmth, because now we have winter, in a word, everything is upside down." The letter is postmarked June 29. The year is not visible.

I confess that I was at first disappointed, though not truly surprised, that my great-aunt, Dvorah Weisman, was not a Yiddish Mme de Sévigné. Yet here was an unexpected reward. First, the discovery of another point on the map of the Diaspora—a third sister who had settled in Argentina. And then some greater sense, however allusive, of what these people—my people—had left behind, probably in the wake of World War I.

It has been already 8 years that I don't feel well, and hardly a day goes by that I shouldn't feel pain, but everything seemed like good times compared to what we lived through in Russia. We didn't pay attention to the pain and didn't make a fuss over it, therefore, when we came to the golden land, Argentina, I was in big trouble, I constantly fell apart, and a couple of times, I was so sick, that no one believed I'll live through it.

Dvorah's comparison of her bodily pain to the difficulty of life in Russia enchanted me even as it provided a poignant measure of the suffering endured in the world they had left behind. In my hunger for palpable traces of my missing family, I loved hearing, I confess, the litany of her misery and its crude language—my great-aunt's sense of triumph over those who thought she had a belly full of *schmaltz* and not serious trouble that had to be removed. Here was a glimpse of all I would never know directly but only guess at through the accident of a single saved letter.

I was especially captivated by the Yiddish lessons the translator gave me when she included the occasional idiom that defied translation. And "everything helped like a bean tied to the wall," Dvorah commented, describing the useless remedies applied by the doctors before she was finally was taken to a Jewish hospital in Buenos Aires for surgery. It's not hard to see why the letter was saved, with its vivid description of suffering and triumph, near death, emergence into life—Dvorah adds Khaye ("life," in Hebrew) to her name—and a Passover story, as she dates the narrative. On the back of the envelope, someone has penciled a short list in Yiddish (blanket, nightgown, dress) and jotted down numbers that appear to be an addition of expenses. Were these items to be sent to Argentina? Preparations for a trip? Perhaps the letter circulated for a while, to be shared with friends and relatives, until finally set aside for safekeeping.

The letter seemed both strange and familiar. I had a sense of recognition, but a recognition that was almost impersonal: the Yiddish cadences that still echoed from my childhood—and from generations of Jewish comedians. (I was to learn that there were two options in translating from Yiddish: one, this one, to make the English read like Yiddish, through a slightly convoluted syntax and mixture of tones; the other, a less inflected English, where the cadences of the mother tongue are domesticated.) My great-aunt's prose was animated by a cosmic *oy* of the kind even my parents mocked. We were quite deliberately not *Oy vey* Jews. At best, Dvorah resembled the long-suffering female protagonists of Anzia Yezierska's stories, who lived and complained on the streets of the Lower East Side, where my grandparents lived when they arrived in America.

Some time after the translation of the letter, and only by accident, looking for something else, I stumbled upon the letter from my father's cousin in Canada that I mentioned earlier, a letter that briefly recounted Dvorah's life, but whose narrative I had failed to absorb.

> Sadie was my mother's (Sarah's) sister. There was another sister, Dvorah, who was the youngest of the females; she emigrated with her husband and their two daughters to South America. I think only one daughter remained—the parents died, the older daughter was unhappy in her marriage and committed suicide, leaving a small boy. We have lost track of the remaining daughter.
>
> I know nothing about where exactly they came from, but I do know that Sadie's father was the overseer of an estate belonging to some wealthy Russian man. As for Raphael's [my grandfather's] family, we know nothing about them.
>
> My mother took me and my sister Fredi to visit our grandparents once when we were very young children—too young to remember very much about the visit. It was the one and only time we saw them.
>
> We, my parents and my three sisters and I, emigrated to Canada in 1921. My mother corresponded with her parents and family, but they never saw each other again. However, Sadie visited with our family many times and my mother went to New York once, I believe.

They never saw each other again.

On my second reading, I was riveted by this geography of loss. Three sisters. A suicide. This was Chekhov territory. How could I have missed this? I suddenly remembered a photograph, with Yiddish writing on the back, of a family that, like the letter from Argentina, felt both strange and familiar. Maybe this was a picture of Dvorah; she looked something like my grandmother, with the same square, unhappy face. I turned again to translators. The inscription, in the neat script I now imagine I recognize, reads: "Dear Sister! Don't be surprised that I have aged so. It is still me and it is all for the good considering that I am constantly ill. And the *naches* we are experiencing from our daughter demonstrates I am made of iron. So this is why I look so ugly."

The third sister, Dvorah, whose face reflects the unhappiness of the message, holds a little boy, presumably her grandson, on her lap. I'm guessing that the young woman standing behind the bench is the boy's mother, the *naches*-giving daughter. This young woman resembles her Canadian cousin to a striking degree. The translators have left the word *naches*—the mixture of pleasure and pride that come typically from children—untranslated. It's one of the dozen or so Yiddish words I know, that the translators figured I would know. But this message does not completely make sense: how can *naches* explain that Dvorah is made of iron, and, at the same time, that she is ugly?

But I queried my translators on this and they said that *naches* had to be understood as maternal sarcasm—that it was the daughter's bad behavior (*tsuris*, the opposite of *naches*) that was causing the suffering and the ugly face.

My grandmother Sadie emigrated to the United States in 1906, her sister Sarah emigrated to Canada from Russia in 1921, and Dvorah emigrated to Argentina, probably around the same time, as part of the emigration waves out of Russia after World War I. It seems unlikely that Sarah and Dvorah ever saw each other again. But Sadie saved her youngest sister's letter—a letter addressed to her older son—and now, through translation, that trace of their relation at least has been found, along with the photograph.

Some time around the miraculous recovery described in the letter, but probably close to the time the snapshot was taken, Dvorah's daughter, "unhappy in her marriage, committed suicide, leaving a small boy." That makes for a very sad story. I wonder why my father never told it, or my mother, if she knew it, which I imagine she did. Is it perverse to wish to have heard suicide stories when I was growing up? It would have interested me to know that my adolescent sorrows had a genealogy. Maybe it would have made me feel, like the narrator of Maxine Hong Kingston's *The Woman Warrior*, that I too had a "weeping ghost" in the family repertoire to redeem me through my own stories. Or an antecedent for my depression.

Not long after I pieced this story together, my sister, another great saver, showed me a wedding picture she had come across in an album belonging to my grandmother. We must have looked through the album right after my father's death but, recognizing no one but my grandparents, put it away for safekeeping, along with the rest of our mysterious archive.

The portrait was taken in Rosario, and on the back of it there was a greeting in Yiddish: "Dear brother-in-law, sister, and Kipnis family! Enclosed is Etyushele's wedding picture. Please let us know that you've received it. The second picture is for Sam and his family." By then I knew who the sender had to be. The mother, my great-aunt, whose daughter's marriage had ended badly—and with it, her life.

Recently my sister surprised me with a story about my father that I had never heard. One morning on her way to school, she said, a pigeon shat on her shoulder. The bird shit—or "doo doo," as she reconstructed the anecdote—was so copious that she returned home to change her blouse. This made her late for school, and so my father, who kept lawyer's hours and never left for the office before nine-thirty, wrote her a note, "Please excuse Ronna's lateness. She was besmirched by a pigeon." I laughed because "besmirched" was so typical of my father's choice of words, what a friend who comes from a family like mine calls "lawish"—the distinctive diction of Jewish-American lawyers whose childhood world spoke Yiddish. Lurking behind the hypercorrection of *besmirched* was the vulgar accuracy of *beshissen*.

But I've interrupted my anecdote. Before he gave my sister the note for the teacher, my father recalled an episode from his childhood, a day when he had been late for school because he had been sick and throwing up. His older brother, Sam, who emigrated as a Yiddish speaker when he was nine, composed the following sentence: "Please excuse Louis's lateness; he vomult." From "vomult" to "besmirched" within a single generation! The linguistic traces of assimilation continue to echo long past the silences about what was left behind. I felt a pang of envy because my father hadn't told me the story, and a wave of sadness for all the stories from our past that would never get told, even in translation.

NOTE

Versions of this essay have been published in *Hotel Amerika* 7, no. 2 (Spring 2009), and in *What They Saved: Pieces of a Jewish Past* (Lincoln: University of Nebraska Press, 2011).

|||

LEGISLATING
INTIMACY

Women's

Work,

State

Control,

and the

Politics

of

Reputation

9

"SECURITY MOMS" IN TWENTY-FIRST-CENTURY U.S.A.
THE GENDER OF SECURITY IN NEOLIBERALISM

Inderpal Grewal

The issue of security/insecurity that pervades the United States at the beginning of the twenty-first century has led us to think about feminisms in newly urgent ways. How do we understand what is happening with feminism when feminist discourses are used to bomb and to liberate, when feminist discourses, strategies, and injuries become available in new and unintended ways to empower, to secure, and to destroy? While most "security" expertise addresses questions of states and geopolitics while ignoring gender, race, or sexuality, many feminist scholars have brought in these categories usefully for a critique of masculinity and militarism and by linking feminism with peace, victimhood, and innocence.[1] But feminists also have discussed the ways in which women and feminists as well have participated in nationalism and militarism or the ways in which domestic ideologies have supported national and imperial goals.[2] To add to this conversation, I want to discuss the security state not in terms of the victimization of female subjects by a militant masculinity but rather to see how militant masculinity within neoliberal contexts rearticulates the public/ private division that has consequences for feminist and female subjects and citizens.

The particular female citizen-subject that is the focus of this essay is the American figure called the "security mom"—a subject called forth in the 2004 presidential election by Republicans seeking election on the platform of security against "global terrorism." This subject reveals how the neoliberal state maintains and disavows its powers and limits through the dynamic of public and private. While some critiques of neoliberalism suggest that it incorporates the reduction of the state and increasing power to private realms,[3] for feminists neoliberalism has come to mean new collaborations and dynamism between public and private patriarchies. It is the flexibility and malleability of these entities, with their shifting borders, that characterize contemporary gender practices.

In the 2004 national election, one issue of gender that made its way into the newspapers was the existence of the new category of voters called "security moms." While after the election of Barack Obama, Sarah Palin might be a prime example of this figure, I want to turn to the beginnings of this particular phase of militant motherhood in national politics in the United States after 9/11 and the 2004 presidential election. The manifesto of the security mom appeared in *USA Today* (20 august 2004), written by the self-indentified conservative syndicated columnist Michelle Malkin.[4] She tells us in her document that she owns a gun, votes, is married, and has two children. From her Web site we see she is also Asian American and calls herself a Generation Xer.[5] Here is what she writes: "Nothing matters more to me right now than the safety of my home and the survival of my homeland. . . . I am a citizen of the United States, not the United Nations." Since 9/11, she writes, she has begun to monitor everyone around her:

> I have studied the faces on the FBI's most wanted terrorists list. When I ride the train, I watch for suspicious packages in empty seats. When I am on the highways, I pay attention to larger trucks and tankers. I make my husband take his cell phone with him everywhere. . . . We have educated our 4 year old daughter about Osama Bin Laden and Saddam Hussein. She knows there are bad men in the world trying to kill Americans everywhere. This isn't living in fear. This is living with reality. We drive defensively. Now, we must live defensively too.

She quotes "conservative activist Kay R. Daly, a security mom of two in Northern Virginia," who says, "Hell hath no fury like a momma protecting her babies."

The two figures she fears most: "Islamic terrorists" and "criminal illegal aliens." She claims she will vote for whoever provides the most security: "Do they have what it takes to keep suicide bombers off our shores and out of our malls?" She ends by saying: "To paraphrase the Iron Lady, Margaret Thatcher: Gentlemen: this is no time to go warm and fuzzy. Security moms will never forget that toddlers and schoolchildren were incinerated in the hijacked planes. . . . As they (the terrorists) plot our death and destruction, these enemies will not be won over by either hair-sprayed liberalism or bleeding-heart conservatism."

There is much to take apart here, at the very least the reliance on children as innocent victims; the reference to Margaret Thatcher—a strong proponent of neoliberalism; the nationalism in which home is joined to homeland and motherhood is about protection by the state; the notion of risk in the United States, in malls and inside the state; the surveillance of everyone around her, including her husband; the education of her daughter; the articulation of a nationalist motherhood project; and the designation of liberals as "hair-sprayed" or conservatism as "bleeding-heart," which bizarrely and deliberately mixes up political ideologies (perhaps ridiculing any remnants of George W. Bush's "compassionate conservatism") for an election-year spin.

There is really little point in seeing such campaign politics as a reflection of a real society "out there." While certainly the security mom is a neoliberal subject that is a product of particular conservative discourses, we cannot ask if the security mom was "real" or not, which was the response from liberal commentators. For instance, Ellen Goodman responded to the Malkin argument by claiming that the security mom was an urban legend rather than a new demographic, while stating that women are intrinsically concerned about security.[6] In her column "The Myth of 'Security Moms'," Goodman writes: "I'm not denying women's concerns about security. Whether it's domestic violence or crime in the streets or terrorism after 9/11, women are more likely to worry that they or theirs are vulnerable." For Goodman, women have a more complex notion of security that "includes healthcare security, retirement security, and economic security." For Goodman too, who identifies as a liberal, the term "security" could be assimilated to the naturalized anxiety of a group that called itself "women." While "security mom" designated women as mothers seeking to protect their innocent children, a figure that is not so new in the history of modern nationalisms, or even American nationalisms and racisms, Goodman allied herself with the U.N. internationalist human security paradigm of the 1990s also as an essential aspect of women.

While these are two different feminist subjects, there is much they have in common, and certainly the claims of women's essential relation to security as protection would be one such element. Though Malkin would certainly distance herself from Goodman's liberalism, both rely on feminist discourses in which private and public are brought together, though Malkin's nationalism and Goodman's internationalism also set them apart. Yet Malkin's "security mom" relies on feminist discourses as its condition of possibility; feminism is part of the assemblage of the "security mom" though Malkin's neoliberal conservatism reworks and rearticulates many of its other discourses. What is thus clear is that feminism has become so much a part of both liberalism and neoliberal discourses of the state and its limits and so intrinsic to an American nationalism that conservative women have to utilize it even while they distance themselves from particular politics of this expansive agenda.

How to explain such subjects in the twenty-first century that bring together a nationalism that produces women as mothers, a conservative feminism, and new forms of racialization and deracialization (Malkin, for instance, is an Asian American)? Here we see ideas of security and safety within imperial and transnational contexts of neoliberalism and geopolitics, and the production of a subject in relation to rearticulations of outsider and insider, home and homeland. If the question is whether the security mom is a new subject or an older national subject, we can see that it is a new subject that is created by links to the foundation of historically embedded raced and gendered subjects. This subject becomes visible as an assemblage of disparate and connected forms of power, new strategies, and technologies, as well as deeply sedimented American notions of home and domesticity, imperialism, and feminism.[7]

While the relationship between empire and domestic ideologies has a long history in the United States, showing indeed that public and private were integrated and often interconnected, neoliberalism emphasizes the fluidity between the two realms, rather than the decline of the state or the triumph of the private realm. In the public realm of defense, the state remains powerful and utilizes female subjects within the private sphere, such as the mother, to produce soldiers and patriots, as well as to become both the subject and agent of security through new surveillance technologies that emphasize the governmentality of security. Right-wing religious movements as nonprofit organizations have become increasingly powerful and able to control debates, policies, and practices of reproduction, marriage, and family, resulting in greater emphasis on the proper role of women as mothers. The state collaborates with religious organizations, especially supported by conservative politicians such as George W. Bush, to control reproductive rights in the name of the heteronormative Christian family. At the same time, neoliberalism suggests that the state is unable to provide security, and thus it disavows its ability to protect all citizens. While "women" as subjects cannot be wholly protected, the mother becomes increasingly the target of security if only because of the state's push toward militarism as its main task. The work of security is governmentalized through the function of the mother, who marks therefore both the limits of the liberal state and the rise of religious organizations as the arm of the state and as its accomplices in the private realm of the family.[8] The neoliberal state allies with nonprofit organizations to move its work of security into spaces that are deemed to be private in multiple ways. Thus this militant mother defines the proper-gendered female subject within the home, the community, civil society, and the nation. Such privatization makes the personal into the political and the work of security as the job of the heterosexual family. In the process, neoliberalism in the United States has come to invest the mother-subject as being powerful and essential to the privatization of welfare and security.

In an insightful essay on the media representations of child sexual abuse in the United Kingdom during the year 2000, Vicki Bell argues that these representations produced the mother as a vigilante and the abuser as a monster in order to avert the crisis of security and legitimation for the government and its institutions.[9] She suggests that the protests by mothers demanding more information about the whereabouts of those accused of being pedophiles constituted challenges to the political rationalities of neoliberal governments. Thus she states that "neoliberal government runs smoothly only if parents can trust that the state is indeed providing both a basic level of general security and trustworthy information by which to make their risk assessments." Because the state was unable to provide information that would enable mothers to "fulfill their roles as rational risk assessing parents," the popular media supported government in reaffirming the state as rational and the mothers as emotional and nonrational.

While Bell's essay is extremely useful in analyzing the role of nationalism and the limit and crisis of neoliberal security, and the ways in which feminist arguments were ignored by the state, the media, and mothers, the context of the United States and its neoliberalism is somewhat different. First of all, there was no outcry in the media that these mothers were vigilantes or even unreasonable in their quest for security. Second, in the context of the "war on terror," these mothers saw themselves as allies of government as well as antagonists (similar to the situation in the United Kingdom) demanding more security from the government. Third, since the state is viewed with suspicion by conservatives while it is also asked to provide security, neoliberal governmentality meant that the mothers in the United States were much more willing to see themselves assisting the state in doing the work of surveillance. In the terms that Michelle Malkin uses, mothers were ready to step in to support the masculine, heteronormative state; their challenge to the state was to ask it to be more heteromasculine, and more heteronormative. Because of the powerful role of nonprofit conservative Christian organizations in support of the heterosexual family and the mother, such mothers were the allies of the state rather than its challengers.

The most productive framework for understanding this assemblage of the mother as proper citizen is Foucault's ideas of power and technologies of rule and subjectivity, in which modernity involves the production of the self-actualizing and self-improving subject though forms of biopower and biopolitics.[10] Foucault's analysis of biopolitics argues that forms of governmentality emerged in the production of modern bodies (as in the production of demographic groups such as "security moms"). So the questions that feminist scholars must grapple with are the following: What makes this constituency? What notions of governance emerge to seek security as an attribute of motherhood? What ideas of motherhood produce a gendered group called "women" in the United States? How do what Malkin calls "Islamic fanatics" and "criminal illegal aliens" become key figures for producing this mother in early-twenty-first America? And what role does the geopolitics of the United States state play in the production of this subject?

To answer these questions, we have to turn to the connection between forms of power, both sovereign and biopolitical, and their relation to geopolitics. Such a connection has not been at the center of most recent debates within cultural theory, but certain aspects of this have been discussed by feminist postcolonial scholars examining the "woman question" within European colonial politics.[11] However, within theorizations of biopolitics, this has come to be a critical relationship. According to Foucault, around the seventeenth century, the population comes into existence as a collective whose life has to be governed and regulated. While biopower is that name given to disciplinary power operating on bodies, biopolitics is the exercise of power over populations. Foucault distinguishes biopolitics and biopower from pastoral power and sovereign power, although he also suggests they can exist in combination. Biopolitics is the power over "man as species" and the ability to "make live," while the exercise of sovereign power is to "take life and let live." Biopolitics is distinguishable from, although it also permeates, sovereign power that can "take life or let live." It deals with the "population as political problem, as a problem that is at once scientific and political, as a biological problem and as power's problem."[12] Thus biopower and biopolitics are two aspects of a joined form of regulatory power, the concern for "making live" coming together with mechanisms for regulating populations as collectives. Governmentality is understood as a "variety of ways of problematizing and acting on individual and collective conduct in the name of certain objectives which do not have the State as their origin or point of reference."[13]

Foucault reveals that sovereign power and biopolitics permeate each other. The connection is necessary for either to function because it is only through this permeation that we understand the regulatory mechanism of security. Foucault himself points out that an apparatus such as health security may necessitate at some point the decision about who can be made to live and who can be let to die—Foucault sees here also a point at which racism enters but also a point at which the state has its limits because it is unable to provide full security to all. Though different, these mechanisms, it can be argued, are often inseparable from the point at which one cannot exist without the other. Regulation works to push into the realm of the biopolitical the decision about which groups can be let die, where the "let die" and the "make die" cannot be distinguished, or where "making live" or "denying life" merge.

To turn to the way in which bio- and geopolitics come to work together, the security mom reveals the interaction between sovereign power, biopower, and biopolitics. By making the mother into both the subject and the agent of security, motherhood becomes governmentalized. However, the increasing power of the religious right and the control of reproduction suggest that the mother is also the target of sovereign and disciplinary power producing domestic subject-citizens whose empowerment coincides with the needs of the nation and state. It is precisely through this production of privatized spaces of security that we can examine Malkin's security mom, as security reworks space and territory in an assemblage in which gender, race, and nation become integrated; security enables domestic space to expand rather than simply contract, resulting in the production of national and imperial subjects. Thus for middle-class, mostly white women, suburbanization is expanding rather than encapsulating, since in security, the state and the private work together to extend biopower into the realm of biopolitics, where self-protection and the mother's protection of the family become part of governmentality.

In rearticulating the connection between public and private within the contemporary discourse of security, we see that biopower and biopolitics also exist in relation to geopolitics, since security is produced through citizens-subjects who are differentiated from noncitizen-subjects—such as the undocumented immigrant, the racialized Other, and the foreign "terrorist." If geopolitics can be understood as a politics that relies on the use of space and territory within the politics of security,[14] as well as the production of national/imperial subjects through the noncitizen-subject, it also includes the incorporation of nationalisms that underlie the definitions of state and citizen. It is nation—that horizontal and vertical relationship that brings the "people" together with the individual—that connects disciplinary power with population and geopolitics. Gender and race are key here, as race and territory, home and homeland, family and nation become coextensive. Security becomes not just a project of the state or individual subject within liberalism that relies on civil society, but of a nation as a space of security that is both deterritorializing and reterritorializing.

For Foucault, security was a mode of liberal power. Colin Gordon has argued that for Foucault, liberalism becomes an "effective practice of security" that is the "political method" of modern governmental rationality.[15] Security is directed at the ensemble of the population. As Foucault states, "War is the motor behind institutions and order. In the smallest of its cogs, peace is waging a secret war."[16] It is this war of power that produces marginalization, which Foucault calls "race" since it is about the limits of security and the ability of liberal states to provide unlimited security to all people. To a certain extent, Foucault suggests a program of study for this problem rather than a definite answer; given that his work does not go beyond the European liberal democratic state, there is much work to do here. In this spirit, it is important to understand the "security mom" as a locus of power where the neoliberal state pushes at the nation to do what it no longer wishes to take on or is faced with its own limits and thus shifts the work of biopower and biopolitics onto the nation. The nation can individualize (produce an identity that is flexible and changing) as well as collectivize in a biopolitical sense; in the form of the nation-state it pervades public and private realms, civil society as well as state institutions. It is the medium through which forms of power can permeate each other or become indistinguishable. Especially in neoliberal contexts, the nation is critical in demarcating citizens from noncitizens, in producing zones of public and private that are differentiated as well as fluid and dynamic.

To return to Malkin, it is difficult to understand the "security mom" simply through an understanding of the state. Rather, in seeing America as empire and nation and the "security" mom as a key citizen-subject of that empire, one is able to see how the nation produces the context of security in which bio- and geopolitics become inseparable; governmentality works to produce subjects for whom both are interconnected. Thus I would like to focus on Malkin and the security mom in order to make three main points about gender of security at the present time. First, home and homeland become coextensive by reworking notions of public and private, especially within discourses of dangers to the family, and especially in discourses concerning domestic violence and sexual predators. Second, the interconnections between the war on terror, hyperconsumption, suburbanization, and new spaces of "community" help to dissolve the difference between public and private space. Thus, contrary to arguments that suburbs are insular and closed in, they also expand out and colonize. Third, the racial and gendered politics of saving and killing are at work within security and visible in the neoliberal governmentality that recuperates motherhood. And fourth, what is critical to the gendering of security is the ways in which a feminist critique of the family has become used for the protection of the family.

The relation between home and homeland and the relation between the nation and the domestic space has long been part of the imperial history of the United States.[17] In the contemporary phase of the American empire, the articulation between security, safety, and gender is used to call upon this historically sedimented discourse of home and homeland to create the "security mom." Motherhood has been a key subject for militant nationalism in many parts of the world.[18] In the U.S. context, in addition to the historical discourse of domesticity and security, new laws and second-wave feminisms have provided new technologies of motherhood. In the case of Malkin, this self-identified Generation Xer "security mom" was earlier the "soccer mom," that demographic of suburban women from the 2000 election. What is interesting here, of course, is that the soccer mom became one only after the passing of Title IX and its resulting inclusion of women in athletic programs in schools and colleges. Title IX of the Educational Amendments of 1972 is the landmark legislation that bans sex discrimination in schools, whether it be in academics or athletics. The opening up of these opportunities for women and girls in schools and colleges resulted in a huge influx of women into recreational as well as school and college athletic opportunities for girls and women, especially in middle-classes suburbs, and the "moms," as they are called, use huge SUVs for transporting their children and teams to various games and tournaments. The need to transport girls (and boys) from one suburb to another remains the task of middle-class mothers, many of whom fear venturing outside the suburb.

Thus the biopower that produces forms of gender through Title IX and girls' sports in the United States, works along with regulatory mechanisms for mobilizing parental "concerns" for the welfare of children. These mechanisms include keeping girls healthy and thin through sports activities, keeping them "out of trouble" (as any sexual activity is called), and preparing them for college. Among "parenting experts," parents, and media, there circulate daily alarms regarding how dangerous the world now is for young girls and boys. There are reports that teenagers are sexually more active than any other generation, that menstruation onset has moved to even earlier age, that a host of drugs are corrupting the youth. From a variety of political sources there are suggestions that there are crises of "self-esteem," drug use, eating disorders, body mutilation, and suicides. While there are some of the same considerations for boys, the discourse around young girls is alarmist and crisis-laden, resulting in a moral panic that is productive of parents and mothers as protecting subjects. Combined with the suburbanization of motherhood through life in gated communities or suburbs (for the safety of children), driving very large SUVs, and forms of super-consumption in the name of the family, such fears have come to regulate middle-class, normatively white motherhood in the latest version of the American dream.

Given the suburbanization of the middle-class in the United States, such alarms and parenting panics have spatial dimensions, just as laws addressing and responding to these panics come to have technological and spatial dimensions. The emergence of gated communities have privatized public and community spaces, but what is important also to note in the discourse of violence by Michelle Malkin is that her home and the mall become coextensive spaces; the shopping mall has become an extension of the domestic space, and indeed, the feminization of this space has enabled it to become coextensive with the home. Domestic space has expanded, incorporating the suburb, the gated community, and the shopping mall. Indeed, all of these are now policed to prevent any one seen as a "stranger" from entering. For youth, the mall remains one of few places to congregate outside the home proper and to escape parental supervision, though it is monitored extensively by private security firms and policemen hired by retailers and mall owners.

The technologization of surveillance has created new realms of security that have become key to proper parenthood. Internet surveillance is now central to this set of fears, becoming both a tool and a danger. While Megan's Law allows parents to identify spaces where "sexual offenders" reside and thus presumably to be able to move away from those areas (if one is financially able to do such a thing), it embodies also a set of dangers (as does TV) that brings a fear of strangers as intruders into the private space of home. "Media experts" warn parents about sexual predators lurking on Web sites and luring girls to meet them outside the home. Web sites and social media used by the young are seen as particularly dangerous; parents are warned not to let children give personal information on the Web or to identify themselves in ways that might lead these predators to them. The Internet is thus viewed as both essential to learning and as full of dangers, with sexual predators visiting every Web site. Finally, "Amber Alerts" move the discourse of dangers and sexual predators even to the freeways, suggesting that mobility and travel are connected to dangers to children. In all of these cases, the most probable and likely and proximate sources of danger are effaced in favor of fear of strangers.

In addition, surveillance works on the body rather than being only in the biopolitical realm, and it becomes productive of the American "civic body" of the mother, as Minoo Moallem terms this embodiment of the nation and state.[19] Whereas historically the discourse of safety from rape for white women justified the lynching or imprisonment of black males, middle-class women's (not just white here, since Malkin is Asian-American) safety required the detention of Muslim males as well as surveillance of everyone. Malkin's discourse on proper motherhood attempts to erase racial difference in favor of national belonging through participation in the governmentality of surveillance. Thus Malkin states she practices "surveillance" on her husband for his safety, never letting him leave home without his phone. It is also a widespread belief that all children today, especially teenagers, must have a phone so that they can be reached and so they are "safe." Even the "family" must be under surveillance, since in this neoconservative imaginary it is under threat (from gays/lesbians/ feminists/terrorists, etc, according to conservative and right-wing religious commentators) and must be constantly supported. The discourse of safety, sometimes thought of as a more nostalgic moment of pastoral power, provides the governmental apparatus through which surveillance become widespread, the family is seen to be threatened, and technologies of individual empowerment become inseparable from those of surveillance.

If this constant surveillance creates a continued war of terror within public and private spaces in the United States, there also emerged at the end of the twentieth century an articulation of female and feminist anxiety over the safety of women and of violence against women. Creating community service and nonprofit agencies within the United States and outside it as well, feminist groups have taken action against patriarchal violence by working both with and outside the state and making domestic violence into a highly visible issue. The topic of domestic violence has become a staple of the media, local and national, with great regularity, almost every day. Often some exasperated reader/viewer will bemoan such coverage, asking why the media pays attention to such cases rather than to the War in Iraq or to Washington or to international politics. Yet these cases have come to be an important aspect of the biopolitics of terror in constructing the "domestic space" and its characteristics. While the main political message of those struggling to identify and end domestic violence was to argue that women were more at risk for physical assault at home rather than outside the home, and from relatives, lovers, and acquaintances rather than from strangers, the War on Terror has curiously inverted this belief as well as relied upon it to create a gendered form of widespread anxiety about safety that takes shape in ideas of motherhood and family. At the same time, however, the continued reporting of this family violence suggests that neither the state nor private and feminist agencies can completely eradicate this violence, however much they may try. Indeed, the continuation of domestic violence and family violence also sutures feminist projects to state projects.

The discourse of security in the war on terror suggests that the greatest danger is from the outside, from aliens, and from non-Christians, especially Muslims (the address is, of course, to other Christians, thus putting Muslims outside the nation). Waco, Oklahoma City, right-wing white terrorists in the United States—are all now figures of the past. What remains is not just the fear of the terrorist other but also the fear of what is called "the underclass," "gangs," and "drive-by shootings" that constitutes the realm of disciplinary and sovereign power that pervades urban life in the United States. This fear of the racialized figures of male violence of course hearkens to a history of such representations, particularly of the black or brown male rapist, figures that are once again called forth when Malkin states her fear of "Islamic fanatics" and "criminal illegal aliens"—racialized figures that infiltrate the so-called "safe" space of her home.

This anxiety about security relies upon fears of safety called upon by the pervasive discourse of domestic violence but inverts the politics of domestic violence by suggesting not that home is a site of patriarchal violence but that the home must be protected against it—thus taking on nationalist discourses of protection by externalizing danger. Safety became gendered through the discourse of domestic violence and the war on drugs (remember that), through which astounding rates of incarceration of racial and gendered bodies, understood to be more violent than the white, heteronormative male subject, came to be normalized. A number of cases of "stranger rape"—of white women by non-white men—came to be publicized in the press; drugs were seen as a scourge that contributed to the formation of gated communities in urban locations, even while saying "no to drugs" programs in schools were produced and continue to be part of school curricula, suggesting that all children were susceptible to taking drugs and must be protected.

Given the war on terror's insistence upon the "foreign" as the greatest threat to the domestic space, even remedies against domestic violence have come to be quite problematic. As the movement against domestic violence came to be accepted as state policy under the Clinton Administration with the Violence Against Women Act of 1994,[20] the state stepped into the work against domestic violence. State protection, however, came through increased policing as well as through private and public programs that directly helped women. In the Bush administration and under Attorney General John Ashcroft, work on domestic violence continued as an extension of initiatives to preserve "family values" and enable women to pass on values to their children. Initiatives on more DNA testing and the 2003 President's Family Justice Center Initiative[21] continue to show us how the initial feminist critique of the family as a site of violence has come to be resignified as a site for the reestablishment of family values through the use of "faith-based" remedies and the reestablishment of the power of the head of the family/the nation to protect. The state and family collaborate in such projects, not only through supporting and lobbying for legislation and programs to "save" heterosexual marriage but also through augmenting the role of mothers through such programs as homeschooling.

In the most recent history of the Violence Against Women Act, reauthorized in 2005 and signed into law by Bush in January 2006,[22] there was an increase in the programs directed toward youth and children, with special protections and service provisions in the areas of sexual assault services and direct services for children. One provision, entitled Strengthening America's Families by Violence Prevention,[23] also targets youth and children as specially to be protected against violence; another is the predictable legislation against trafficking, especially of children. While there are some useful and necessary services provided in this legislation, there is also a major emphasis on the collection of the DNA of anyone arrested or detained, even if they have not been convicted, especially if the person is not a U.S. citizen.[24] In general, the legislation also strengthens law enforcement and information gathering. While some may think that the Act's inclusion of provisions for the Border Violence Task Force and the National Gang Intelligence Center might be extraneous to the main agenda,[25] the reliance of the security mom on the threats of so-called aliens, terrorists, and urban gangs are very much part of the security assemblage that produces the national/imperial gendered subject.

It is in this constant move to displace the violence of the family onto what are described by Malkin as "Islamic fanatics," "criminal illegal aliens," and "urban gangs" that we see how biopolitics functions only with geopolitics. The technologies of welfare by the state that are directed at preventing violence against women and promoting the safety of women have become part of a project to resuscitate patriarchal and pastoral power in its disciplinary aspect not only by the mobilization of a law-and-order apparatus or through new bureaucratized means of surveillance (expanding DNA banks), but also by externalizing danger and moving it onto bodies seen to be a foreign threat to the nation or to proper citizens.

Finally, it is precisely by expanding the domestic space into the extraterritorial realms of American power (Guantanamo, Iraq, Afghanistan) that a denial of "family" violence can take place. At the same time, it is through the daily reminder of the domestic as a violent space that the boundaries of "home" become unstable. This instability and constant tension between the daily reminders of "private" violence in the home and the attempts to displace this violence onto dark and foreign others suggests that a feminist critique of neoliberalism must see the public/private divide as dynamic and fluid. Feminists cannot simply yearn for a return to the state or see the state as a protector of the "commons" that exists outside of power. Especially in the imperial context of the United States, where sources of violence are seen to be nonwhite males and where particularly non-Christian cultures are understood, within almost two centuries of knowledge production, as intrinsically violent, anti-women, and antifeminist, it is important to insist that for many women, it is the heteronormative, middle-class family that can also be a source of violence.

What are apparent in the discourse of security in the United States in the early twenty-first century are the limits of security, even for the normative and proper subject of the nation, the mother. Domestic violence makes this insecurity visible, even though conservatives struggle against it in the name of the family and feminists struggle against it in the name of women. The limit of the state's power itself becomes a form of regulation, in which biopower and biopolitics work together to produce "mothers" as necessary to the protection of the family of Americans and to require them to participate in their own surveillance for their security.

NOTES

My thanks to Caren Kaplan and her colleagues and students at the Cultural Studies Forum at the University of California, Davis, for feedback on this paper. And also to the terrific "Think Again" Group for their very useful comments. Also to audiences at the University of California, Irvine, Critical Theory Institute and Purdue University.

1. The tradition of feminism that connects women to pacifism and patriarchy to war is visible even before second-wave feminism. Virginia Woolf in *Three Guineas* makes this connection and argues for pacifism. Feminists from Betty Freidan to Vandana Shiva have continued in this tradition.
2. See, e.g., Claudia Koonz, *Mothers in the Fatherland*.
3. See, e.g., David Harvey, *A Brief History of Neoliberalism*.

4. Michelle Malkin, *USA Today*, 20 August 2004. All Malkin quotes are from this opinion piece.
5. Malkin's Web site is available at www.michellemalkin.com.
6. Ellen Goodman, "The Myth of 'Security Moms.'"
7. See, e.g., Laura Wexler, *Tender Violence*; Amy Kaplan, *The Anarchy of Empire in the Making of U.S. Culture*; Vincente Rafael, *White Love*.
8. John Micklethwait and Adrian Wooldridge, *The Right Nation*; Sara Diamond, *Not by Politics Alone*.
9. Vicki Bell, "The Vigilant(e) Parent and the Paedophile."
10. Michel Foucault, *The History of Sexuality*.
11. See, e.g., Tani Barlow, *Formations of Colonial Modernity* or the essays in Kumkum Sangari and Sudesh Vaid, eds., *Recasting Women*.
12. Michel Foucault, "*Society Must Be Defended*," 243–47.
13. Michel Foucault, *Power*, 375–76.
14. Gearóid O'Tuathail, *Critical Geopolitics*.
15. Colin Gordon, "Governmental Rationality."
16. Foucault, "*Society Must Be Defended*," 50–51.
17. See, e.g., Wexler, *Tender Violence*; Kaplan, *The Anarchy of Empire*; Rafael, *White Love*.
18. Sara Ruddick, *Maternal Thinking*.
19. Minoo Moallem, *Between Warrior Brother and Veiled Sister*.
20. The Violence Against Women Act of 1994 (VAWA), Pub.L. 103–322, 108 Stat. 1902. http://thomas.loc.gov/cgi-bin/query/z?c103:H.R.3355.ENR:
21. "The President's Family Justice Center Initiative," U.S. Department of Justice, www.ovw.usdoj.gov/docs/family_justice_center_overview_12_07.pdf.
22. The Violence Against Women and Department of Justice Reauthorization Act of 2005, Pub. L. 109–162, January 5, 2006, 119 Stat. 2960. Available at: http://www.govtrack.us/congress/billtext.xpd?bill=h109-3171.
23. Ibid., sections 1006–1007.
24. Ibid., sections 1001–1004.
25. Ibid., sections 1006–1007.

BIBLIOGRAPHY

Barlow, Tani. *Formations of Colonial Modernity*. Durham: Duke University Press, 1997.
Bell, Vicki. "The Vigilant(e) Parent and the Paedophile." *Feminist Theory* 3, no. 1 (2002): 83–102.
Diamond, Sara. *Not by Politics Alone: The Enduring Influence of the Christian Right*. New York: Guilford Press, 2000.
Foucault, Michel. *Power. Essential Works of Foucault 1954–1984*. Vol. 3. Ed. James Faubian. New York: The New Press, 2000.

——. *"Society Must Be Defended": Lectures at the College De France 1975–76*. Trans. David Macey. New York: Picador, 2003.

——. *The History of Sexuality*, vol. 1. New York: Pantheon, 1978.

Goodman, Ellen. "The Myth of 'Security Moms.'" *The Boston Globe*, 7 October 2004. http://www.boston.com/news/politics/president/articles/2004/10/07/the_myth_of_security_moms/.

Gordon, Colin. "Governmental Rationality: An Introduction." In Graham Burchell, Colin Gordon, and Peter Miller, eds., *The Foucault Effect: Studies in Governmentality*, 1–52. Chicago: University of Chicago Press, 1991.

Harvey, David. *A Brief History of Neoliberalism*. Oxford: Oxford University Press, 2005.

Kaplan, Amy. *The Anarchy of Empire in the Making of U.S. Culture*. Cambridge: Harvard University Press, 2002.

Koonz, Claudia. *Mothers in the Fatherland: Women, the Family and Nazi Politics*. New York: St. Martin's Press, 1987.

Malkin, Michelle. *USA Today*, 20 August 2004. www/usatoday.com/news/opinion/editorials/2004–07–20-malkin_x.htm (accessed 12 December 2004).

Micklethwait, John, and Adrian Wooldridge. *The Right Nation: Conservative Power in America*. New York: Penguin Books, 2004.

Moallem, Minoo. *Between Warrior Brother and Veiled Sister: Islamic Fundamentalism and the Politics of Patriarchy in Iran*. Berkeley: University of California Press, 2005.

O'Tuathail, Gearóid. *Critical Geopolitics*. Minneapolis: University of Minnesota Press, 1996.

Rafael, Vincente. *White Love: and Other Events in Filipino History*. Durham: Duke University Press, 2000.

Ruddick, Sara. *Maternal Thinking*. New York: Ballantine, 1989.

Sangari, Kumkum, and Sudesh Vaid, eds. *Recasting Women: Essays in Indian Colonial History*. New Delhi: Kali for Women, 1989.

Wexler, Laura. *Tender Violence: Domestic Visions in an Age of U.S. Imperialism*. Chapel Hill: University of North Carolina Press, 2000.

Woolf, Virginia. *Three Guineas*. New York: Harcourt Brace, 1966.

10

"LIKE A FAMILY, BUT NOT QUITE"
EMOTIONAL LABOR AND CINEMATIC
POLITICS OF INTIMACY

Tsung-yi Michelle Huang and Chi-she Li

Driven by the tides of economic globalization, migrant workers have emerged as a salient presence in work forces worldwide. As migrant workers cross national borders to work, they are also confronted by barriers of other kinds. In Taiwan, as is the case in many other countries, legal measures ensure that migrant workers remain transients, unlikely to overstay their term of employment and make claims on precious social resources. In addition to legal borders, symbolic borders proliferate, and they prove as effective at barring migrant workers from substantial participation in society, culture, and politics. As the political philosopher Étienne Balibar reminds us, "Even as the usefulness of borders in civic space is becoming more problematic, one can observe a tendency for collective identities to crystallize around the functions of imaginary protection they fill, a fetishism of their lines and their role in separating 'pure' identities."[1] Within this proliferation of borders, a perplexing paradox emerges: when one draws physically close to and perhaps also grows emotionally close to a migrant worker on the micropersonal level via daily life encounters or media representation, the force of legal and symbolic boundaries is renewed.

We consider this proposition by surveying a number of films created since 2000 in Taiwan. Inspired by a profound concern for the place of foreign laborers in Taiwan, these sympathetic works enjoy increasingly wide distribution, shown in workers' campaign activities, local film festivals, public TV broadcasting, and even government-sponsored screenings, and they play an increasingly significant role in raising the public awareness of the cause of transient domestic/health-care workers from Southeast Asian countries.[2] We will analyze closely three works representative of recent filmic representations about how foreign laborers are treated in Taiwan: *Hospital 8 East Wing* (2006), by the Taiwan International Workers Association (TIWA) labor organization; *Nyonya's Taste of Life* (2007), aired on the Public Television Service (PTS) Channel to promote a multicultural approach to integrating foreign spouses into Taiwanese society; and *We Don't Have a Future Together* (2003), also aired on PTS. All of the three films are well-meaning endeavors that cast migrants into affable cinematic images by likening them to family members. Meanwhile, the audience is encouraged to identify with the protagonist workers and develop cinematic intimacy with the characters. Our assessment of cinematic representation takes place within a critical analysis of the constitutive logic of domestic/health-care work—namely, the paradox of being like a family, but not quite. In spite of the fact that these filmmakers attempt to include the migrant laborers as "one of the family," such benign efforts are entangled with complex border management. As a result, in examining these three films we inquire into the unsettled tensions between a congenial affirmation of migrant workers and the constrictive governance of migrant labor for the state's regulatory purposes.

We read the production and possibilities of intimacy in the context of global labor migration in different ways in relation to the three films under discussion. Specifically, an analysis of the documentary *Hospital 8 East Wing* serves to expose the tension between the family trope and state control of short-term migrant domestic/care workers and the lived ambivalence of the family trope within Taiwanese families. The other two films, *Nyonya's Taste of Life* and *We Don't Have a Future Together*, spin around the romance genre to narrativize the family trope. It is undeniable that romance carries a widely accepted structure, providing viewers with identifiable codes for cultural narration, but we will examine how this genre may obscure or ironically highlight the issues and concerns of migrant workers. In *Nyonya's Taste of Life* we look at how the metaphor of being "like a family" is transformed into a marriage scenario to displace the structural exclusion of foreign laborers onto a matter of personal dislikes. In *We Don't Have a Future Together* we open up the possibilities of cinematic intimacy, informed by Neferti Tadiar's suggestive methodology.[3] Exploring the Filipino subaltern experience under the sway of contemporary capitalism, Tadiar sees in various forms of narrative not only dominant narratives but experiences closer to emergent history. After breaking the presumed connections of narrative apart, one might find fragmentary, discontinued threads of narratives, which lead to a better understanding of how a person is situated delicately on the threshold of being subjugated and liberated. Inspired by Tadiar, we attempt to read *We Don't Have a Future Together* for what lies beyond "being like a family but not quite." De-centering the marriage plot of the film, one sees a possibility of salvaging cinematic intimacy from being caught up in the exclusion of migrant workers.

To set the films in context, as the population ages[4] and social welfare has been proven to be inadequate, Taiwan has accelerated its importation of migrant workers and become heavily reliant upon foreign labor. A steadily increasing number of foreign domestic workers is an absolute social necessity.[5] This can be understood within Foucault's analysis of biopower in modern societies as functioning to "foster life or disallow it to the point of death," through the management of the life of populations in general.[6] Numbers, statistics, segmentation, and the classification of populations are central to biopolitical management. Cultivation of vital activities and the reduction of risks such as disease and aging to the population are preoccupations of those who govern.[7] Biopolitical governmentality in contemporary neoliberal societies subsumes social relations under the economic rationality of costs and benefits and delegates a large part of the management of society risks to the individual. One is assumed to make use of market alternatives to maintain and "develop" oneself. State-sponsored social welfare mechanisms have been cut back and private markets sustain life functions.[8] A country like Taiwan depends on foreign laborers to sustain the functions of the family and the health and lives of family members. When the national welfare system or a large number of families are unable to sufficiently provide for unproductive citizens with unproductive bodies, the labor of migrant workers is needed. In this way, the work of migrants is an important tool in the state's management of life.

In a receiving state such as Taiwan, short-term migrant workers assume the now largely commodified and privatized work of caring for children, the elderly, the sick, and the disabled in order to maintain the proper functioning of the family and to release women into the waged labor market. Under such circumstances, migrant workers have been asked to instrumentalize their own various biological and psychological capacities in the service of others. The migrant workers are used to "foster life" in the population of the receiving state but are restricted by state labor regulations from developing their own abilities and emotional capacities beyond the interests of their employers. This critical understanding of the biopolitics of using foreign domestic/health-care workers in Taiwan enables us to comprehend the constitutive logic of domestic/health-care work, which takes the paradoxical form of being "like a family but not quite."

Just-like-family rhetoric circulates in this context of the global marketization of intimate care work. A number of researchers have discussed the negative effects of this rhetoric for migrant workers. Even if "just-like-family" provides a framework for negotiating the inward tensions between the commodification of domestic labor and the intimacy purchased,[9] it simultaneously elides tensions and problems between the employer and the laborer.[10] The just-like-family rhetoric can be a programmed demand, and the job performance of the foreign laborer is often measured against the standard of familial devotion. "Like-a-family" is predicated on the concealed lack of reciprocity between employer and domestic laborer and the assumption that filial relations are one-sided and unreciprocated. To account for this one-way pressure on migrant workers, we suggest an analytical term, *the instrumentalization of life*, an unalienable aspect of the biopolitical governance of foreign domestic/health-care workers, to delineate the demand for migrants' emotional attachment.

1

Hospital 8 East Wing is an unusual cultural product for Taiwan. Commissioned by the Taiwan International Workers Association, this documentary speaks to the interests of migrants as workers. The director, Hui-Zhen Huang, hoping to debunk stereotypes of foreign migrant workers,[11] presents the story of how a health-care worker from the Philippines named Lisa establishes a quasi-filial affinity with her client, Grandpa Ren. The director also utilizes a cinema verité style, which de-emphasizes the obtrusiveness of the camera and editing, presumably to allow the represented to unfold their stories by themselves. The film shows Lisa working caringly as a quasi family member and in this way suggests that migrant workers deserve the audience's respect for their reliability in helping maintain the functioning of Taiwanese families. Yet the director's intentional promotion of workers' rights sometimes jars with the chosen style of presentation, so much so that toward the end of the film we find glaring inconsistencies. Despite the film's use of Lisa's genuine emotional attachment to a Taiwanese client as the basis of Taiwanese empathy with and understanding of migrant workers, the strict demand imposed on the migrant workers, as defined from the viewpoint of the state/employers, still simultaneously intrudes violently to place migrant workers outside of the family.

Hospital 8 East Wing realistically captures the everyday experiences of foreign health-care workers at the Renai Municipal Hospital in Taipei. The opening shots unveil a difficult working environment: inside the ward we find mostly paralyzed old men who can hardly speak. As Grandpa Ren is turned onto his side, the shot shows that he is wearing a diaper. Patting his back to ease his discomfort, Lisa speaks softly, "Grandpa, I can barely move you. You are so heavy." Another shot displays a Vietnamese worker, A-Ying, helping Grandpa Wei turn over, cleansing his body and massaging his arms. Then, Lori from the Philippines, attending to Grandpa Chou, gently stroking his head, spells out the anxiety she felt when first charged with the care of an invalid patient: "While I was new here, I was very nervous because I had no experience. I cried every day. There were so many things that I didn't know. Especially in the face of a patient like this, I was truly nervous, two hundred percent nervous. . . . I didn't want anything to go wrong with him. So I got nervous. I am hoping he could hang on, but I'm no God. Ha! Ha!"

Confined in the hospital wards, these foreign laborers are always on call, carrying out such tedious and even demoralizing tasks as suctioning, back patting, hand and foot massaging, changing diapers, and cleaning bed sores. The careful tending to the patients' bodies, captured by the camera, attests to the intimate and physical nature of the work. As Carol Wolkowitz describes it, such work "takes the body as its immediate site of labour, involving intimate, messy contact with the body, its orifices or products through touch or close proximity."[12]

However, the candid style of the film also allows the audience to observe the complicated meaning of the nonphysical labor, commonly termed *emotional labor*.[13] By definition, emotional labor requires the worker to invoke their emotions, consciously or unconsciously, or otherwise to repress them in the course of performing their duties.[14] Undertaking emotional labor in a context in which illness and death are the stuff of the everyday raises particular challenges for workers. This emotional investment may be rationalized by the laboring subject as part and parcel of the various roles played—imagining oneself as a loving mother allays the negative questioning of self-esteem that plagues many domestic workers who contribute additional physical and emotional labor in exchange for greater respect, dignity, and self-regard.[15] Sometimes emotional labor is psychologically therapeutic, allowing domestic workers to transfer their attachment to, and responsibility for, their own families to those for whom they care.[16] These observations of emotional labor are primarily premised on the laborer's initiative, or positive feedback from the working environment. Nevertheless, in responding to the needs of those in their care, domestic workers and health-care workers are not simply fulfilling the duties of employees, but also are expected to play multiple familial roles commonly performed by a mother, a daughter, or a wife.[17]

Examining this TIWA documentary, we are also obliged to question how an employer demands emotional care in terms of quasi-filial relationships.

It is noteworthy that in *Hospital 8 East Wing* the favorable images of migrant workers are partly derived from their attentiveness to their clients in a manner similar to the emotional attachment a family member might exhibit. The camera constantly evokes empathy from the audience through the laborer Lisa, who is cast in the image of a family member. For example, Lisa speaks of Grandpa Ren as "my grandpa" with profound sincerity. She says: "Since my grandpa started to get worse, I haven't slept well. . . . You know I cried. I have been crying, not because I am afraid to lose my job but because I am sad and I love Grandpa. I have been looking after him for almost four years." This guileless presentation of Lisa bespeaks cinematic appreciation of how she is devoted to her job. However, in presenting the moments in which the worker is close to the client, as a family member might be, the camera might also surprisingly create a situation in which viewers are encouraged to endorse an employer-approved quasi-"filial" devotion through narrative and cinematography.

Hospital 8 East Wing portrays an extreme case of emotional investment in one's labor in the form of bereavement. Toward the end of the film, we witness a care worker's emotional burden on clear display. At Grandpa Ren's funeral, Lisa, the "surrogate granddaughter," weeps like a family member (figure 10.1). The film concludes with a series of shots of Lisa. It shows first an empty ward, the appearance of a tearful Lisa at the memorial service, Grandpa's family taking her to a final viewing of the deceased, and Lisa sobbing and covering her face. What Lisa offers at the funeral is not merely labor, but total emotional involvement. In spite of the director's cinematic advocacy of the cause of workers, when the cinema verité style collapses the representation of Lisa as a laboring family member and the pathos of her mourning at the funeral into one scene, the intractable ambivalence of the meaning of Lisa's emotional labor looms large. The foreign laborer's emotional attachment to the one she cares for may enable her to maintain a great degree of self-esteem or alleviate homesickness and guilt, as the case of Lisa shows; however, in the words of Dodson and Zincavage, such emotional attachment might also turn the laborer into a "prisoner of love." Moreover, although one could find in Lisa's sorrows the expression of self-willed devotion, ironically, for the same reason, it could be argued that the emotional investment of the worker is naturalized in tune with the interests of the state and the employer. Seen in this light, the cinematic intimacy established in the shots of Lisa's mourning might also perpetuate a demand for the laborer's emotional labor.

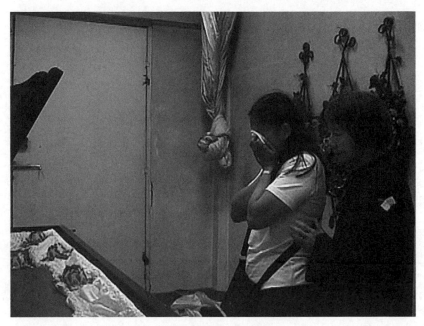

10.1 Crying like a family: Lisa at the final viewing of Grandpa Ren
(Courtesy of Taiwan International Workers' Association)

The rift in meaning in the rhetoric of like-a-family erupts when the funeral is over.[18] In this last scene Lisa stands apart from the family. But in the shooting and editing of the very final scene, both the employer and the director were put to the test regarding how they understand the meaning of the like-a-family appeal. The first take of the family photo did not include Lisa (figure 10.2), but there was a second take, at the request of Grandpa Ren's son, Lisa's employer, upon seeing the film crew's camera and Lisa's isolation after the initial photo was taken (figure 10.3). Lisa was immediately invited to be incorporated into the family photo. This subtle uncertainty surrounding the film finale, which marks the closure of a migrant worker's contract with this family, is intriguing. In the discarded second take, the employer reveals a sentiment of familial inclusion. The employer's quick afterthought, considering the family photo without Lisa not good enough, provides a glimpse into the complexity of the like-a-family rhetoric in practice. When the family trope is used to account for the intimacy between the employer's family and the migrant worker, the employer is stranded in the ongoing, unresolved ambivalence of the migrant worker's simultaneous exclusion from and inclusion in the family.[19]

Regardless of the employer's attempt to include Lisa, the first take was chosen, in which Lisa was left to the side, standing alone, waiting with plastic bags in hand, while the others join in the family photo. Huang explains the decision to include the one, but not the other: "The one without Lisa is closer to reality since the employer is not aware of the intervention of the camera."[20] Huang's non-interventionist editing diverges from the previous sympathetic understanding of the situation of a migrant worker. The asserted filial affinity between the employee and the employer ends with the service term. In this film's last shot, we find the migrant worker alone. The conspicuous absence of Lisa in the family photo keenly illustrates the negative side of the constitutive logic of emotional labor: her filiality terminates with the death of her charge.

The inconsistency of the cinematic attitude toward Lisa shown between the funeral scene and that of the family photo-taking allows one to reflect on the ambivalence of the like-a-family behavior and the tensions therewith. In one sense, the respect for Lisa is generated out of recognition of her labor on behalf of the family. But within this intimacy, one nonetheless hovers on reaffirming the border between family and contracted labor.

10.2 Family photograph without Lisa
(Courtesy of Taiwan International Workers' Association)

10.3 Lisa's exclusion from the family photo
(Courtesy of Taiwan International Workers' Association)

2

Nyonya's Taste of Life was aired on the PTS Channel and has been perhaps the best known work about migrant workers and foreign spouses to have appeared in Taiwan, likely due to its warm depictions of foreign domestic workers and a foreign spouse. The director Chih-Yi Wen states that this film is inspired by her memories of an Indonesian worker who looked after her own grandmother.[21] Three foreign laborers—Cindy, Youchai, and Sari—come across one another in a small town. Cindy, from Indonesia, cares for an elderly outpatient with Alzheimer's disease. Sari, also from Indonesia, labors as a domestic worker in the household of her widowed employer; she is torn between the competing courtships of her boss and Youchai, the Thai worker. The film ends with the happy marriage of Sari and her employer. The fates of Sari and Cindy offer two quite different strategies for resolving the unstable ambivalence of the trope of familial inclusion.

Among the three films under examination, this one is the most direct in defining the intimacy between a migrant domestic/health-care worker and her employer in familial terms. Central to *Nyonya's Taste of Life* is the emotional bond of the female migrant workers with the employers' family—a bond so strong that it leads to marriage. Sari turns from a foreign worker into a local wife after a crucial moment for the employer's business. The family restaurant's business has been dropping off drastically since the death of the employer's wife. Occasionally customers come, but the employer can barely cook. Seeing her boss in such an embarrassing situation, Sari tries first to pacify the customers and then swiftly prepares satay skewers for them. Satisfied with the food, customers say to the boss, "You'd better use her menu instead." In consequence, the restaurant is retooled to offer a Southern-Asian menu, with the help of Sari and Cindy. The employer's financial crisis is miraculously solved by Sari's Southern Asian Nyonya food, and the employer decides to marry Sari. The marriage looks like an expected result of her domestic service. Transforming Sari from a migrant worker to a foreign spouse, the film pushes the trope of "just-like-family" to an extreme scenario in which the female protagonist becomes a real family member, as if to resolve the unstable ambivalence noted in relation to the previous film.

The representation of Sari's interracial marriage benefits from the little daughter-in-law (小媳婦 *xiao xifu*) formula so familiar in Taiwanese films. In numerous mainstream melodramas popular in the 1990s, the woman, either newly wed or yet to be married, is mistreated by the husband's family and, above all, by her mother-in-law. The weak, humble, and hardworking woman, nice-looking but by no means glamorous, in spite of obstacles and harsh treatment, persists in her devotion to the family. The husband's mother tends to speak for the patriarchal interests of the family, usually the well-being of the children, and for the social respectability of the men in the family. The position of the audience in relation to the characters in this formula is stable: the innocence of the daughter-in-law and her dedication to the husband's family are beyond doubt, such that the audience always has great sympathy for this female character during her numerous trials and easily grasps the abusive nature of her environment. The Taiwanese melodramatic convention of the "little daughter-in-law" provides a platform for the audience to negotiate with the presence of a foreign laborer, to reckon with the possibility that the worker is not born as a member of our family but could become one. Such a negotiation is perhaps accentuated by the well-accepted practice in which physically or economically disadvantaged citizens in Taiwan tend to regard as a major option the possibility of marrying foreigners from underprivileged regions. Any impasse in Sari's life as a guest worker can thus be explained away in the form of the idiosyncratic antagonism of the mother-in-law(-to-be). Meanwhile, *Nyonya's Taste of Life* appeals to the established pattern of identification in this formula since Sari plays perfectly the role of a docile woman in the employer's family and works for the same patriarchal interests that the mother-in-law staunchly defends.

When the film seeks to persuade the audience to accept a foreign spouse into the imagined community of Taiwan, it speaks to the traditional gender stereotype formed during the time when the three-generation family was the prevailing structure in Taiwan. This use of the little daughter-in-law convention to represent female migrant workers is by no means accidental in the context of Taiwan. For example, one program for social education, a highly notable attempt aired on a major TV channel that was rerun four times, includes interviews with experts and foreign spouses to reveal how a foreign spouse ought to play her role in Taiwanese society.[22] The overwhelming message is that foreign spouses are expected to be true to the gender divisions of the traditional family and are advised to be "submissive" and "self-sacrificing" for the sake of the family. In addition, they are expected to follow a roughly melting-pot version of multiculturalism; they must be capable of displaying their own variety of foreign cultures but mainly follow Taiwanese practices in socialization.[23] The transmutation of the Taiwanese stereotype of a good daughter-in-law occurs in another form as well. With urbanization, the traditional pattern of having a large family—in which three generations live under one roof and those of the middle generation are responsible for the maintenance and production of the old and the junior—cannot hold. Migrant workers are sometimes hired by the daughter-in-law of such a three-generation family to ease her own burdens, and the social expectations of a good wife/nice mother will be transposed onto the hired domestic worker.[24]

The conversion from an ideal domestic worker to an ideal daughter-in-law seems to make the point: the emotional labor of a domestic worker, when performed well, is so genuine that one could see its natural extension into a marriage. This sameness between a migrant worker and foreign spouse reveals much. The logic works both ways. A good worker makes a good wife; conversely, the characteristics of a good wife could be bent backward to be applied to a migrant worker. It is perhaps not surprising that, in spite of the difference between marriage and employment, Sari as a married person dramatizes the specific demands imposed on a foreign domestic worker, including self-sacrifice and the obligation to raise the welfare of the employer's family to the top of her concerns.

This dominant narrative of marriage sutures rifts between the migrant domestic/health-care worker as a family member and outsider. Alternatively, such ambivalence can be displaced onto personal preferences and dislikes. Such a displacement mainly occurs in Cindy's story. The tension between "like a family" and "not quite family" is played out in a sharp and constant contrast set up between the daughter and son of Grandpa Feng. The work performance of Cindy is not much different in nature from the one displayed by Lisa in *Hospital 8 East Wing*. The film elaborates, by means of close-ups and long takes, on the minute attention Cindy pays to the person for whom she works, valorizing a good worker's immeasurable emotional attachment to the family, though it was unwritten in her job description. The sister Fang-Fang is the ideal employer who sees the foreign laborer as almost a family member, while her brother Hsiao-Hao stands at the opposite end of the spectrum, seeing no place for Cindy in his family. Fang-Fang is grateful for Cindy's commitment. On the grandpa's birthday, she gives Cindy one of her old *qipaos* (旗袍) as a token of appreciation. Notably, the gift is seen not as recognizing Cindy's domestic work but rather as a generous gesture to recognize her as a quasi family member. Fang-Fang recounts that when she turned eighteen, her father had given her a tailor-made *qipao*, the traditional one-piece Chinese dress, as a birthday gift every year until she got married.

By giving Cindy a *qipao*, Fang-Fang symbolically continues the family ritual for her father (and possibly intimates a trajectory toward marriage). Yet this reenactment is tainted by a later confrontation with her brother. As soon as Hsiao-Hao sees Cindy wearing his sister's *qipao*, he is enraged and bursts out: "Family? Who is she? What would people think if they see her like this? She is my mom? My wife? Or another sister of mine?" Fang-Fang tries to reason with him: "She takes care of dad every day, spending time with him. She is just like family." The filial assumption infuriates Hsiao-Hao all the more, and he accuses Fang-Fang of "taking sides with an outsider, a servant!" Hsiao-Hao provokes an argument again after Cindy leaves and insists that their father spend his remaining years in a nursing home that can provide proper professional medical care, something that a worker like Cindy is not capable of. The film clearly favors the warmth of Cindy's care over the cold clinical environment. For instance, the moment Fang-Fang says that the old man needs companionship, the film cuts to a shot of the helpless face of the father. These enduring quarrels between daughter and son make it clear that the sister's gratitude is more sensible than the brother's blatant rejection. In other words, the instability of "being not quite family" is reduced to micropersonal attitudes toward migrant workers rather than presented as a structural problem. The film invites the audience to resolve the structural instability by individualizing it and recasting it within the dualism of good and bad employers.

3

The third film, one that we consider more successfully holds in tension the irresolvable structural contradiction built within the just-like-family trope, is entitled *We Don't Have a Future Together*. Directed by Jin-Jie Lin, the film portrays the encounter of a lower-class Taiwanese laborer, A-Yuan, and Erica, a migrant worker from the Philippines. At the beginning, A-Yuan, a young, strong mover, cannot find a suitable marriage partner. His aunt arranges a blind date for him, nagging, "No one wants to marry a mover with dirty hands." After the matchmaking falls through, to whom will a lower-class person such as A-Yuan turn? A-Yuan bumps into Erica, a Filipina who is having difficulty hailing a taxi. A romance between the two workers begins. Poorly educated A-Yuan fantasizes about how he might rescue the beauty from her plight and marry her. However, the reality is that Erica, college-educated in the Philippines but employed as a domestic worker, suffers from discrimination and her employer's sexual abuse in Taiwan, and is subject to the strict surveillance of migrant-worker controls. Due to jagged discrepancies between reality and fantasy, their relationship proves futile. This unhappy ending to an insubstantial romance, in our reading, suspends the transformation between date and mate, migrant worker and foreign spouse; what is more pertinent, it throws into sharp relief the implied continuity between acting like a good family member and a good foreign domestic/health-care worker.[25] In this way, this film occasions an opportunity for not only discovering tensions in direct valorization of the kind of intimacy that visual representations of migrant workers typically act out but, more important, explores the possibility of treating migrant workers as individuals disengaged from the prevailing framework of kinship and intimacy.

According to Jin-Jie Lin, packaging the story of the working-class characters into one of a romance was a strategic decision: "It would be a big step forward if I could foreground just a little bit working-class issues in mainstream TV shows. Yet we still have to meet the tastes of popular culture. So I had to come up with a common denominator, like a popular love story, for the issues that I was about to address."[26]

This romance builds intimacy between the audience and Erica by filming from the point of view of A-Yuan in two ways. First, abundant point-of-view shots of A-Yuan render Erica as an object of desire and infatuation. In the establishing shots, both the narrative and imagery encourage the viewer to identify with A-Yuan as a kind, funny, working-class chap. As a mover, A-Yuan is a social failure with a golden heart; he worships the heroic persona of Bruce Lee for his righteous martial feats and helps underprivileged Taiwanese aboriginals with much-reduced moving rates. Once A-Yuan's good character is established, his gaze continues to mediate between the camera's eye and those of the viewers. The point-of-view shots invite the spectator to experience falling in love with Erica as A-Yuan does, and to see her with affection, if not infatuation. A-Yuan's point-of-view shots bring details of Erica's life into focus—for example, taking a break in the park, taking care of her boss's sick child in the hospital, taking out the garbage, and praying in the church.

The gaze at the migrant's working conditions is also mediated through the romance formula of "boy meeting girl in distress." The harsh working environment of foreign domestic workers is exposed as the archetypal evil for the hero and heroine. Erica's worst "distress" is sexual harassment by her employer. While the act of sexual violence is understated, what happens after she is raped by her employer points to the severity of the crime. First, a long shot of the employer smoking in the foreground of the frame, with only part of the back of his head visible, contrasts with Erica in the background sitting still on the kitchen floor, half-naked. In the following high-contrast shot, Erica, her face indistinct, begins to clean up the mess on the floor. At the same time, offscreen, the employer appeases and bullies her to be quiet about the rape. Agitated by her silence, he yells at her, "Can you hear me? Listen to me. If you tell somebody, will people believe you or me? You are just a Filipina maid." Next is a long shot in which the door bell rings, building suspense into the scene. This shot presents Erica in distress by putting her at the background of the frame, looking small, out of focus, and powerless, and placing the nameless employer in the foreground, telling Erica not to move, with his finger pointing at her. It is not until A-Yuan shows up at the door to look for Erica, like a knight arriving too late to the rescue, that Erica's pain becomes visible to the spectator.

The film's attempt to address workers' issues is subordinated to the conventions of romance. Privileging, implicitly or explicitly, the perspective of a citizen employer or a citizen candidate of marriage, the film ultimately tilts much more toward the story of A-Yuan than that of Erica and renders Erica an object of his gaze and a passive victim.

The director's decision to fuse a dominant popular cinematic form with concerns for representing migrant workers rests on a concern with making the issues accessible through codes or frames of communication that are familiar to the audience. But is it possible to decouple cinematic intimacy from family intimacy and in so doing to denaturalize the demand for migrant workers' emotional investment? This film offers a suggestive possibility for representing migrant workers' emotional labor outside the terms of familial intimacy. The narrative of the film opens up a context in which the viewer can be detached from rather than pulled in by the like-a-family trope.

This film includes Erica's perspective in some evanescent, fragmentary, and undeveloped shots and narratives. The most telling example is the sequence of shots of Erica alone in the park, the scene launched by A-Yuan's point-of-view shot. A long shot traces her walking to a nearby park, sitting on a bench, and flipping open what looks like a notebook. Mellifluous piano music, birds chirping, and the whistle of the wind convey a tranquil moment in Erica's daily life (figure 10.4). The camera then zooms in to show Erica first closing her eyes in a meditative mood and then looking at the trees above. What follows is Erica's point-of-view shot of a beautiful scene of trees rustling in the wind against the blue sky. Unusual in the clichéd language of cinema of romance, this rare moment shows Erica alone, not in relation to others. Here she is neither an object of desire nor a damsel in distress but a worker who manages to find a short break in her daily routine.

Even though eclipsed when viewed from the perspective of the male's romance, other examples also point to Erica's agency, including her managing of or even calculation in the romantic relationship. On their date at the bar, for instance, a Filipina friend advises Erica: "He [A-Yuan] might be a Taiwanese but not well educated at all. Don't lower yourself. You are a college graduate. He's

10.4 A moment of one's own: Erica in the park
(Courtesy of Taiwan Public Television Service)

not good enough for you." Erica's immediate response hints at her possibly belittling judgment of A-Yuan. Instead of defending A-Yuan, she looks at him from a distance as if to confirm her friend's opinion; lowering her head uneasily, she explains, "He is just my friend." This fleeting moment is a typical case in which a migrant worker, like any other transnational, has to convert cultural currencies (here judging A-Yuan's potential mate value by his social capital) based on a personalized exchange rate. If this scene might help us see Erica as a shopper in the marriage market, trying to weigh the choice of her possible mate, the subsequent scenes at A-Yuan's apartment, where Erica refuses his attempts to be physically intimate by saying "you and I are friends," allow us to look at Erica as a woman who is not shy about saying no to unwanted physical intimacy and who is at the same time savvy enough to deal with the sexual tension without hurting A-Yuan.

The "breakup" scene at the end of the film suggests the possibility of showing Erica as independent and strong, capable of reasoning and articulating her thoughts. Erica meets A-Yuan by the river and blames him for beating her boss, which is very likely to cost her job: "Who do you think you are? You are nothing. You are a worker. What if I lose my job? I have to go back to my country. What do I do then?" Before she repeats four times, "We don't have a future together," she expresses her feelings for him and why their relationship would go nowhere: "You are a good man. You have been very nice to me. And I like you. But I have a son already; I need to keep my job. I need my job." Erica is in charge of their relationship, a subject capable of making a rational choice even in a difficult predicament. She is simultaneously embedded within her own and now Taiwanese familial relations.

The image of Erica as a tough transnational worker with her own family relations, capable of rational thinking and decision making—rather than a damsel in distress or an object of desire—stands outside the like-a-family discourse. The potential of the audience's understanding of Erica could be derived not just from the empathy-induced sight of her helplessness, but also from the clearheadedness shown in her own conversion and translation of complex and conflicting demands into a management of her own individual choices and personal gains, a capacity not unfamiliar to a college graduate.

4

At this point, we maintain a dialogue with filmmakers in Taiwan regarding the issue of how to critically understand the migrant worker's labor from the viewpoint of a member of the host country. Through this discussion, we seek to contribute to the overall project of understanding the complex forms of borders mediated and disseminated in cinematic constructions. It is evident that foreign migrant workers are typically represented within familial tropes, even within films attempting a critical stance. *Nyonya's Taste of Life* is the most conventional, with the instability of the trope resolved, either through marriage and literal absorption within the family or through the duality of good and bad employer. For the other two films discussed the situation is more complex. *Hospital 8 East Wing* draws audiences to the plight of migrant health-care workers through an unquestioned and uncritical valorization of their uncompensated emotional labor, even as it documents a foreign caregiver's eventual exclusion in the last scene of the film. Ironically, we find the most critical representation in the least expected genre: a romantic comedy. This is because the domestic worker is shown to have the agency to say "no" to the Taiwanese family, in part because she is already fully absorbed within her own. Each of the films, in different degrees, shows that when migrant workers are recognized, they might not be rid of the possibility of having their life instrumentalized as well. This paradoxical logic of biopolitcal governance—being like a family but not quite—forcefully renders those concerned, including the employer, his/her family members, and even film viewers, caught up in an uncertain oscillation between the compassionate treatment of migrant laborers and the non-negotiated demand of their emotional labor.

NOTES

1. Étienne Balibar, *We, the People of Europe*, 110.
2. Between 2003 and 2008, according to the Documentary Media Worker Union in Taiwan, three documentaries and four fictional films and docudramas on foreign migrant workers were produced. To address the tension between visual intimacy and emotional labor, we find that representations of the relationship between female foreign migrant laborers, especially domestic and health-care workers, better serve our purpose than those of male workers (such as *Shattered Dreams* and *Dreamer*) in that the nature of the working condition of female domestic/health-care workers exemplifies how the sense of intimacy derived from bodily contact and domestic work can be instrumental in representing migrant workers as "like a family." Among the few works that fit the criteria, *Yuning's Return*, though a story of a female domestic worker, is not included because its focus is the family life of the migrant worker back in Indonesia rather than her interactions with local employers.

3. Neferti Tadiar, *Things Fall Away*. We thank Geraldine Pratt for bringing our attention to Tadiar's work.

4. According to the Population Reference Bureau, the total fertility rate (the average number of children that would be born to a woman over her lifetime) in Taiwan was 0.9, the lowest in the world: www.prb.org/DataFinder/Topic/Rankings.aspx?ind=17 (accessed Oct. 23, 2011).

5. Taiwan Association of Family Caregivers, "A View of the Domestic Caretaker's Burden and Its Public Role Through the Foreign Caretaker Policy"; Mei-Chun Liu, "A Critique from Marxist Political Economy on the 'Cheap Foreign Labor' Discourse"; Fang Wang, "Social Welfare System and the Difficulties in Long-term Care." Until 2008, the number of "foreign healthcare workers" or "domestic workers" was more than 160,000, which accounted for 46 percent of the total number of foreign laborers in that year in Taiwan. Even during the financial crisis and economic recession of 2008–2009, a time when the total number of foreign laborers decreased for the first time in ten years, these two types of workers continued to increase (*United Evening News*, Taiwan, 25 February 2009).

6. Michel Foucault, *The History of Sexuality*, 138.

7. Michel Foucault, *Security, Territory, Population*.

8. Meyer, M. H., ed., *Carework*; N. Folbre, *The Invisible Heart*; Ligaya Lindio-McGovern, "Labor Export in the Context of Globalization."

9. Bridget Anderson, *Doing the Dirty Work?*

10. Maruja M. B. Asis et al., "When the Light of the Home Is Abroad," 210. Through the appropriation of kinship rhetoric, domestic migrant workers seem allowed to be part of the family, which in turn might compromise their autonomy and rights and frame their labor as an obligation: Charlene Tung, "The Cost of Caring"; Nancy Folbre, *Invisible Heart*; Rosie Cox and Rekha Narula, "Playing Happy Families"; Ayşe Akalin, "Hired as a Caregiver, Demanded as a Housewife." E.g., the employer might use the rhetoric of family to justify the practice of denying holiday breaks to domestic workers because family members always stay together: Nicole Constable, *Maid to Order in Hong Kong*. In other words, the family rhetoric might be advantageous to the employer, but it is disadvantageous to the rights of laborers: Brenda S. A. Yeoh and Shirlena Huang, "Spaces at the Margins"; Bridget Anderson, *Doing the Dirty Work?*; Pierrette Hondagneu-Sotelo, "Blowups and Other Unhappy Endings"; Nicole Constable, *Maid to Order*; Akalin, "Hired as a Caregiver"; Lisa Dodson and Rebekah M. Zincavage, "'It's Like a Family.'"

11. Interview notes, 22 October 2008.

12. Carol Wolkowitz, "The Social Relations of Body Work."

13. We use the term *emotional labor* instead of *affective labor* because the latter implies skillful control. In the classical definition provided by Hardt and Negri, affective labor means manipulation of "the entire state of life in the entire organism" to facilitate what a job requires one to do (Michael Hardt and Antonio Negri, *Multitude*, 108). Emotional labor, on the other hand, connotes the result of workers' responses not necessarily falling under their cognitive control. We attempt to develop this important implication of emotional labor.

14. Tung, "The Cost of Caring," 69; Carol Wolkowitz, *Bodies at Work*, 77.
15. Asis et al., "When the Light of the Home Is Abroad," 210–11.
16. Romero (1992), 125, quoted in Asis et al., "When the Light of the Home Is Abroad," 211.
17. S. Dyer, L. McDowell, and A. Batnitzky, "Emotional Labour/Body Work," 2034.
18. We thank Hsiao-Chuan Hsia for sharing with us the reception of her students, who find the family photo the most impressive shot of the film. Her students think that Lisa is mistreated—she has done so much for Grandpa, but she is left out of the family photo.
19. We thank both Geraldine Pratt and Victoria Rosner for this insightful point.
20. Interview notes, 11 January 2010.
21. Interview notes, 30 October 2008.
22. "Taiwanese Wives" first aired in 2005 on Chinese Television System.
23. Li-Jung Wang, Wei-Jing Wang, and Shu-Juan Zhu, "The Transnational Community Media in Taiwan."
24. Pei-Chia Lan, *Global Cinderellas*, 124.
25. The director says that originally there were a couple of possible endings: "One [was] to have Erica meeting A-Yuan at the airport before she was sent back home. Another version [ended] with A-Yuan flying to the Philippines to find Erica despite his poor English. The two finally get together" (interview notes, 28 November 2008).
26. Interview notes, 28 November 2008.

BIBLIOGRAPHY

Akalin, Ayşe. "Hired as a Caregiver, Demanded as a Housewife: Becoming a Migrant Domestic Worker in Turkey." *European Journal of Women's Studies* 14, no. 3 (2007): 209–25.

Anderson, Bridget. *Doing the Dirty Work? The Global Politics of Domestic Labour*. London: Zed, 2000.

Asis, Maruja M. B., Shirlene Huang, and Brenda S. A. Yeoh. "When the Light of the Home Is Abroad: Unskilled Female Migration and the Pilipino Family." *Singapore Journal of Tropical Geography* 25, no. 2 (2004): 198–215.

Balibar, Étienne. *We, the People of Europe: Reflections on Transnational Citizenship*. Trans. J. Swenson. Princeton: Princeton University Press, 2004.

Constable, Nicole. *Maid to Order in Hong Kong: Stories of Migrant Workers*. London: Cornell University Press, 2003.

Cox, Rosie, and Rekha Narula. "Playing Happy Families: Rules and Relationships in Au Pair Employing Households in London, England." *Gender, Place and Culture* 10, no. 4 (2003): 333–44.

Dodson, Lisa, and Rebekah M. Zincavage. "'It's Like a Family': Caring Labor, Exploitation, and Race in Nursing Homes." *Gender & Society* 21, no. 6 (2007): 905–28.

Dyer, Sarah, Linda McDowell, and Adina Batnitzky. "Emotional Labour/Body Work: The Caring Labours of Migrants in the UK's National Health Service." *Geoforum* 39 (2008): 2030–38.

Folbre, Nancy. *The Invisible Heart: Economics and Family Values.* New York: New Press, 2001.

Foucault, Michel. *Ethics: Subjectivity and Truth.* Trans. R. Hurley. New York: New Press, 1997.

——. *The History of Sexuality.* Trans. R. Hurley. New York: Pantheon Books, 1978.

——. *Security, Territory, Population: Lectures at the College de France 1977–1978.* Trans. G. Burchell. New York: Palgrave Macmillan, 2007.

Hall, Stuart. "Encoding/Decoding." In Stuart Hall et al., eds., *Culture, Media, Language: Working Papers in Cultural Studies, 1972–79,* 128–38. London: Routledge, 1980.

Hardt, Michael, and Antonio Negri. *Multitude: War and Democracy in the Age of Empire.* New York: Penguin, 2004.

Hondagneu-Sotelo, Pierrette. "Blowups and Other Unhappy Endings." In Barbara Ehrenreich and Arlie R. Hochschild, eds., *Global Woman: Nannies, Maids, and Sex Workers in the New Economy,* 55–69. New York: Metropolitan/Owl Books, 2003.

Hospital 8 East Wing. Directed by Hui-Zhen Huang. Taiwan International Workers Association, 2006.

Lan, Pei-Chia. *Global Cinderellas: Migrant Domestics and Newly Rich Employers in Taiwan.* Durham: Duke University Press, 2005.

Lindio-McGovern, Ligaya. "Labor Export in the Context of Globalization: The Experience of Filipino Domestic Workers in Rome." *International Sociology* 18, no. 3 (2003): 513–34.

Liu, Mei-Chun. "A Critique from Marxist Political Economy on the 'Cheap Foreign Labor' Discourse." *Taiwan: A Radical Quarterly in Social Studies* 38 (2000): 59–90.

Meyer, M. H., ed. *Carework: Gender, Labor, and the Welfare State.* New York: Routledge, 2000.

Nyonya's Taste of Life. Directed by Chi-Yi Wen. Taiwan Public Television Service, 2007.

Romero, Mary. *Maid in the U.S.A.* London: Routledge, 1992.

Tadiar, Neferti. *Things Fall Away: Philippine Historical Experience and the Makings of Globalization.* Durham: Duke University Press, 2009.

Taiwan Association of Family Caregivers. "A View of the Domestic Caretaker's Burden and Its Public Role Through the Foreign Caretaker Policy." *Grey Hair Era* 28 (2006). www.elderly-welfare.org.tw (accessed 8 March 2009); this material is in Chinese.

Taiwanese Wives. Produced by W. Y. Peng and C. Chien. Chinese Television System (CTS), 2006.

Tung, Charlene. "The Cost of Caring: The Social Reproductive Labor of Filipina Live-in Home Healthcaregivers." *Frontiers: A Journal of Women Studies* 21, no. 1/2 (2000): 61–82.

Wang, Fang. "Social Welfare System and the Difficulties in Long-term Care: An Introspection on the Problems of Foreign Care-Takers." *Employment and Training* 97 (2008). www2.evta. gov.tw/safe/docs/safe95/userplane/half_year_display.asp?menu_id=3&submenu_id=464&ap_id=633 (accessed 10 May 2009).

Wang, Li-Jung, Wei-Jing Wang and Shu-Juan Zhu. "The Transnational Community Media in Taiwan: A Case Study of the Program of 'Taiwanese Wives.'" *Chinese Journal of Communication Research* 14 (2008): 267–313.

We Don't Have a Future Together. Directed by Jin-Jie Lin. Taiwan Public Television Service, 2003.

Wolkowitz, Carol. *Bodies at Work*. London: Sage, 2006.

———. "The Social Relations of Body Work." *Work, Employment and Society* 16, no. 3 (2002): 497–510.

Yeoh, Brenda S. A., and Shirlena Huang. "Spaces at the Margins: Migrant Domestic Workers and the Development of Civil Society in Singapore." *Environment and Planning A* 31 (1999): 1149–67.

11

WHAT WE WOMEN TALK ABOUT WHEN WE TALK ABOUT INTERRACIAL LOVE

Min Jin Lee

I got the awkward question at a book club in Westchester.

After the tasty potluck supper, the members and I were sitting in the host's beautiful living room, balancing plates of chocolate cake and mugs of coffee on our laps. It was the spring of 2008; my first novel had come out in paperback, and I was visiting a large book club near Scarsdale. The members were working moms—hospital administrators, teachers, and small business owners—who had somehow found the time to read a debut novel about a Korean American immigrant community based in New York City. The women were second-, third-, and fourth-generation immigrants from Italy, Germany, Poland, Greece, and Russia; they were ethnically white. The members were thoughtful, generous, and lovely. I wanted them to like me.

Most of their questions about the book were softballs. One of the members brought up the characters Ella and Ted. Ella Shim is the beautiful Korean American doctor's daughter married to Ted Kim, another Korean American who works as an investment banker. Ted leaves Ella for Delia, a sexy white woman with whom he feels a stronger emotional connection than with Ella, his wife who is too prissy for him. The estranged Ella later falls in love with a white man named David Greene.

The group consensus was that Ted was a jerk and the women felt sorry for Ella. They thought it was fair that she'd found David Greene—my nice development director character who gets along with his mother.

Then the same member asked: "So what is it exactly with white guys having a thing for Asian women?"

I opened my mouth to speak then closed it.

Trying to help me, I think, she added, "I mean, there are a lot of these marriages in Westchester." Her tone didn't express that this was a bad thing necessarily, rather that she was noticing a phenomenon.

Another member spoke up: "It's not like these girls had a choice. They had to become mail-order brides." Her shoulders sagged with sympathy.

I realized she was serious.

Another woman chimed in, "There's one of those at my church."

"Do you know this for a fact?" I asked quietly.

"No. I don't know them—" She looked at me strangely, wondering why I'd question something so obvious.

"Then how do you know that she is—?" I couldn't even say it. "You don't think that all interracial marriages between Asian women and white men are the mail-order kind?"

"Well, of course," two women said at once.

"Oh." I took a deep breath. Huh.

"Aren't they?"

"No. Most are not."

When I was growing up in the 1970s and 1980s, interracial relationships were rare even in New York City, but in 2008 they seemed almost commonplace in many parts of the country. That said, ubiquity does not make for intimacy: these relationships may have become more noticeable, but that didn't mean that the majority population knew what was going on in them.

I had already spent three hours with the book club. The members were the kind of neighborly women you'd ask to pick up your kid if you had an emergency at work, or the ones who'd bring you a pan of lasagna if someone got sick. These were nice women: every single one of them would have sympathized with any woman from any country who would have had to marry to escape poverty or political persecution.

But it was also clear that they could not have known many women of Asian descent socially. My sisters and friends who had married white men had not been ordered from a catalog. They were American citizens—either native born or naturalized—and just like these women in the book club, they were educated, working mothers who were juggling the balance of their lives. It should not have seemed unreasonable to presume that theirs were also love marriages with the attendant conventions of traditional courtship and engagement and not deserving the stigma of having a financially negotiated marital bond.

My childhood in America had been modest, and I had grown up in a blue-collar neighborhood in Queens, New York. I didn't want to be an out-of-touch elitist jerk putting down mail-order marriages—the very term itself likely to offend its participants—but y'all, this conversation was freaking me out. In my racially and socioeconomically diverse, liberal and, well, politically correct private world, it had never occurred to me that a non-Asian Western person would view all interracial marriages in 2008 as the mail-order variety.

Did these obviously intelligent women view all Asians in America as recent arrivals and foreign? Was it not possible that an Asian woman (from America or elsewhere) could have fallen in love with a white man and vice-versa? What about Yoko Ono, Connie Chung, Wendy Lee Gramm, and Vera Wang? These famous women were married to white guys; none of them would see herself as an empowered poverty survivor nor a rescued unwanted daughter.

Their certitude—that all interracial relationships between Asian women and white men were financially induced—confessed that these otherwise switched-on women had formulated a curiously satisfying narrative to explain this rising trend: the Asian women were third-world victims, and the white men were the kind who had no choice but to go outside the country to procure a foreign wife because they could not get one from home. This, I thought, was an ingenious argument to avoid a conscious racist attitude toward Asian women and to deflect the possibility that some white men may romantically prefer Asian women. Their sympathy was sincere, but it was also a narrative of condescension mingled with defensiveness. They were certain—it was evident in their tone and expression; but even the laziest investigation would have revealed that this wasn't so. Nevertheless, the white women of the book club may have believed this because it diverted them from the possible truth that Asian women (or women of Asian descent) and white men had chosen each other out of mutual affection—no different than the way women and men of the same tribe choose to mate—causing a disturbance to the global pyramid of female desirability where white women have always been on top. Maybe for the kind-hearted white woman, it might be morally preferable to pity the chosen-over-her-ethnically-Asian woman rather than to hate her as a rival.

The women of the living room knew that I was not married to a white guy, so they had felt safe asking me for insider information.

In my unspectacular dating history, I have dated a few whites and Koreans in equal number. When I was nineteen, I briefly dated a white East Asian Studies major who dated only Asian women—a "rice chaser," "Asiaphile," "Asian-fetishist," "yellow-fever carrier," what-have-you. It was not my finest hour. It is nothing short of morally perverse if not intellectually deficient to desire a romantic partner only for his or her racial stereotype; in the same way it would be moronic to want to date white guys exclusively because you believed that all white men are Hollywood good-looking and world leaders. My lame excuse: I'd been dumped by my high school boyfriend of three plus years. I attribute this lapse in judgment to a self-esteem deficit and a wish for an artificial romantic boost. A prolonged diet can make the sugar deprived reach for Nutrasweet.

He was not a bad person, decent-looking and not unpopular, but the experience was regrettable for me nonetheless. He was arrogant, slick, and, not to mention, not my physical type, so it ended within a month or so, and he moved on to the next Asian woman. You love and learn.

Later, when I was in law school, I met my husband Christopher. I was twenty-two. When I was twenty-four, we got married. Christopher had a white American father and a Japanese mother. He was biracial and bicultural. Born in Kobe, Japan, Christopher spoke Japanese at home and grew up in Tokyo, Hong Kong, and America.

My in-laws married in 1963 when antimiscegenation laws were in force in parts of the United States. When they met, my father-in-law, Chuck, was the American Consul in Yokohama, and my Japanese mother-in-law, Matsuko, was the daughter of Count Chuji Kabayama. The Kabayama family fraternized with the imperial household; the Kabayama men graduated from Deerfield, Princeton, and Amherst, and my Japanese mother-in-law attended the elite Shipley School in Pennsylvania when it was single-sex—all of this highly unusual even for a Japanese family from the peerage. My mother-in-law, who died in 2006 from heart disease, was an elegant and creative person, and on occasion, she could be an insufferable snob. My mother-in-law and I were not close.

After the book club meeting, I could not imagine how many times Matsuko must have been viewed as a bar girl redeemed by my father-in-law, the son of an insurance executive from Davenport, Iowa. I wondered if that's why she had name-dropped her famous relations so incessantly.

Not that it's ever possible to know what people actually think about you, but the world provides clues.

For decades, anti-Asian graffiti have been appearing in college bathrooms around the country accusing Asian girls of being promiscuous and boyfriend-stealing. On the Internet, anonymous netizenry routinely call Asian women hos, sluts, and bitches. Film and television routinely depict Asian women and Asian American women as conniving, shrewish, materialistic dragon ladies, or depraved prostitutes and crude sex performers; they are the minor and occasional characters dropped into a scene for villainy or comic relief to contrast against a virtuous white main character.

In response to the high rate of Asian American college admissions, U.C.L.A. is also known as University of Caucasians Lost among Asians, U.C.-Irvine is University of Chinese Immigrants, and M.I.T. is called Made In Taiwan. All of this implies that there are just too darn many Asians everywhere. Again, that yellow peril.

Newspapers chronicle how Asian girls and women from third-world countries are sold, bartered, and denied literacy. Visual media depict economically better-off Asian and Asian American women as hypersexual, forever foreign, social-climbing, fan-snapping tai tais, and as dastardly and cartoonish as pirates. In the minds of the American majority, immigrants from Asia stay immigrants generations after their citizenship papers have been certified; after two hundred plus years of settling down and Westernizing, with only marginal cultural recognition and acceptance to show for it, I wonder if it is possible for the majority to ever see anyone with an Asian face as truly American. One can hope, but the writing on the wall is literally ugly.

Why do media, cultural, and social representations matter? Well, we know that in the absence of intimate relationships—that is, interracial friendships— these shallow images supplant the complex reality of twenty-first-century America.

So, we women—white, brown, and yellow—glance at each other in suspicion, avoid the truth, and miss opportunities, and for what?

My family and I moved to Tokyo from our home in New York City in the fall of 2007. We rented a place in an apartment building for expatriates.

For the first few months, the white women in the building wouldn't say hello to me. They always said hello to each other. I had lived in apartment buildings all my life. One was not often chummy with all the neighbors, but in an elevator or taking out the trash by the incinerator, you said hello to people you didn't know.

As an Asian person in Japan, it was as if I were wearing local camouflage, and visually I'd lost my American citizenship; the Japanese thought I was stupid for not knowing the language, and I was brushed aside socially right along with the majority by English-speaking women from the States and the U.K. My Asian American friend, also an expatriate in Hong Kong, calls this "Asian-blindness" and "Asian-amnesia," because non-Asian expatriate women stubbornly refuse to greet her and won't remember meeting her in social situations, although of course they have.

On the streets, I'd see a white woman greet another white woman as if they'd been reunited with a long-lost family member, and I'd feel bizarrely sad and left out. No doubt, they were feeling as culturally lost as I was, but their racial envelopes communicated their foreignness to each other immediately. For me, I'd have to talk and sound American before the non-Japanese might recognize me as an American or as an English speaker.

There were very few Asian women in the building, and if there was an Asian woman, she was usually Japanese and married to a white man from Australia, Europe, or America. At my son's international school, the Japanese mothers had their own P.T.A., and the white foreign mothers huddled separately. Later, when I met some Korean mothers from South Korea, they made it clear that I was not really Korean, but American. Okay. Expatriate Japan was beginning to feel like some sort of elective social apartheid.

So I made an effort. At the minimum, I would say hello to anyone I saw in the building. I overcame whatever issues I had and would greet anyone first. One evening, after some chatter about the weather in the elevator, an African American male resident told my son and me: "It sounds like you've spent a lot of time in America, because you have an American accent."

"We are Americans," I said, trying not to sound pissed off, but I was. Because African Americans experienced the nonsense of wrong assumptions (a.k.a. racism) routinely, I was holding him to a higher standard of sensitivity. However, he and I were riding in an elevator in an Asian country that prides itself (inaccurately) on its mono-racial population and pure blood culture. The current governor of Tokyo, Shintaro Ishihara, has called homosexuality abnormal, Africans unintelligent, women beyond childbearing years useless, and all foreigners including those born in Japan, namely Korean-Japanese and Chinese-Japanese, for several generations the source of crime in Japan. In a nation where prime ministers have a shorter shelf life than a carton of milk, Governor Ishihara is wildly popular in its largest and most important city; he is likely to run for his fourth term in office and win. The man in the elevator, my neighbor, made an innocent mistake, yet my feelings were hurt; our world was just so confusing.

Months later, there was some media coverage about my book, and for a few people I was known as the Korean American novelist. Gradually, I made more friends. In social events, I was usually the only Asian woman in the room. Japanese women married to foreigners were rarely invited to social events. They had their own social lives with other Japanese women married to white men. I kept seeing parallel universes with only the white men crossing among them. In this odd expatriate world, white men from America and Europe were like old-fashioned colonizers or explorers—having boundless access and reach to frontiers both foreign and private. Manifest Destiny called upon the young man in the nineteenth century to go West; in the twenty-first century a young man of gumption and wanderlust may trek as far east as Asia—but where are we women to go, or do we end up with whoever comes courting?

I learned about Charisma Man—a decade-old cartoon strip created by Canadian Larry Rodney and initially published in a now-defunct expatriate magazine, *The Alien*. Charisma Man is a white male superhero who is a financial and social nobody back home, but when he steps off the plane in Japan, he transforms into a sexy powerhouse—a virtual magnet for Japanese females. The punch line? His chief nemesis and kryptonite: Western Women. In their presence, Charisma Man returns to his country-of-origin identity: loser. It appears that Charisma Man's creator, Larry Rodney, was commenting on the absurd amount of extra credit he received for his white maleness. Charisma Man as a concept is funny and clever, but it also problematizes interracial relationships, indicting Japanese women and white men for being blind to their respective social realities. The cartoon narrative is hinged on the fantasy projections each group has toward the other, thereby deeming Asian interracial relationships as unreal and ultimately suspect. Taken to its logical conclusion, the de facto normal and real relationships are intraracial ones. So is Charisma Man an antimiscegenation narrative?

One afternoon, a couple of weeks before summer break, I was invited to a luncheon by a woman at her home. There were five white expatriate wives and myself. I rearranged my work schedule and attended happily.

The host, a talented cook, made a gorgeous curry, and I had seconds.

The women were attractive and fun stay-at-home mothers, and their lives sounded glamorous to me. Like other well-off expatriates I'd met in Tokyo, they owned summer homes in England, France, Hawaii, Long Island, and New England. The day after school let out for the summer, they'd board a flight with their kids and leave Japan for two months, while their husbands remained in Tokyo.

The host mentioned that she'd spent the morning consoling a friend whose husband had been sleeping with a Japanese woman from his office.

The women nodded, having heard this story before.

"In the summer when we're gone, some of the boys go to the bars and make out with these slutty Japanese girls," one said. "My husband saw our neighbor down the hall sucking on a Japanese woman's tit at a club the other night, and his wife has no idea."

"When I'm gone, my husband plays a lot of golf," said another.

Another cooed reassuringly, "Your husband would never—"

"But you know, some of these Japanese women are so aggressive. They don't have the same morality that we do. They don't care if a man is married. It's all fair game to them."

More nods.

They knew I was Korean American, and like most expatriates in Tokyo, they knew enough about the Japanese occupation of Korea and how the Korean-Japanese (*Zainichi*) people are discriminated against in Japan. The women at the table felt safe dishing about Japanese women, never thinking that it could offend me.

My head was about to explode. The stereotype went that Asian women are submissive, and now Asian women are sexually aggressive, and white husbands are prey? Japanese women are amoral?

Being an Asian American feminist means that, like it or not, I have a pan-Asian political identity even when I know pan-Asia does not exist, especially on the ground of an Asian country.

From 1910 to 1945, Japan brutally occupied Korea—true; and Japan continues to treat third-generation Korean-Japanese who are essentially culturally Japanese as if they are foreigners without equal protection under the law—also true. The Japanese did evil, unspeakable things during the Pacific War to many Asian nations, and none of them have forgotten the atrocities except for the Japanese history books that still bizarrely paint Japan as the victim.

Nevertheless, we Americans know enough Asian American history: Vincent Chin, antimiscegenation laws, the internment of the Japanese during World War II, the Chinese Exclusion Acts, and mob lynchings of Chinese railroad workers and miners—the list continues.

As an American and as a Korean American, I am very glad Japan lost the war, very glad; but I am not the only one to wonder if we Americans would have deployed atomic weapons against a white enemy?

I live in one of these transnational quandaries that are rife with the privilege and problems of belonging to a number of tribes: American, Asian American, Korean American, immigrant, Christian, feminist, and now expatriate.

When I arrived in Tokyo, there were several pointed instances when my half-Japanese husband witnessed Japanese women being hostile to me either because I was not Japanese or because I was ethnically Korean, and those instances made me realize that the Japanese nationals did not have a pan-Asian feminist identity.

That said, the majority of modern Western women, especially young girls, rarely identify as feminist, so sometimes I feel like I'm dressed like a suffragette holding a placard that says ERA NOW at the outset of the third millennium, because the global truth is that it's still lousy for most women around the world.

The women at the luncheon continued to talk about married men who behaved badly with the local women. I knew the host only a little because our sons played together. I did not know the others. At the table, I hadn't defended Japanese women, and I felt sort of sick.

When the dessert came out, I begged off, making excuses about work.

Not much later, an article in the *Times of London* came out about the exodus of rich Western bankers in Tokyo and how that was affecting the gold-digging Japanese women who "shark" at Roppongi Hills for them. After two decades of the declining Japanese economy, Japanese nail salon girls could no longer find Australian sugar daddies. The tone was satirical, but the joke on Japanese women was nothing to laugh at. You could have replaced "army base" for the fancy expatriate bar called Heartland and "prostitutes" with young Japanese women who longed to shop with their Western banker boyfriends' credit cards, and you'd have a story about sad hookers after a military base decamps. The article was forwarded to everyone I knew, and they all chuckled at how right the author had got it.

Wherever I went, the article was discussed, although it made Japanese women sound like amoral whores who would sleep with anyone who would buy them an Hermes handbag or the equivalent. The women at the luncheon now had an article, from the *Times of London* no less, to bolster their argument that all Japanese women who'd married bankers were materialistic desperadoes who had closed the deal.

Of course, we didn't talk about how many white women from the sex industry—dancers (strippers) and club hostesses (occasional prostitutes)—had married into the wealthy expatriate class. In my time in Tokyo, I had met working-class white women from "pink-collar" professions—flight attendants, waitresses, preschool teachers—who had married up. No one questioned the romantic veracity of these relationships—*Pretty Woman* had already removed that stigma. When a banker married a model—a predictable and ordinary liaison—was it a money-for-looks status trade? Maybe. Maybe not. Such unions were not under scrutiny at this luncheon, because when people of the same race marry for money, we allege a higher purpose. As the *Pretty Woman* soundtrack suggests, "It Must Have Been Love."

If interracial relationships were inherently suspect, then it was possible for the Western expatriate wife to call Japanese women amoral and opportunistic, because Japan is the world's third-largest economy, and Japanese women were not starving. The Westchester book club women had felt sorry for the women in interracial marriages, because they thought those poor Asian girls had to marry out to escape starvation. So sex for food is pitiable, but sex for $6,000 purses made you a scheming bitch.

It's worth noting that these are the mainstream narratives of interracial relationships, but they are not the narratives from the Asian women themselves.

Because I am me, I worry about lots of things that are not literally my problem, and the fact that Asian women are skewered by the media, reviled by well-off, expatriate white women, dismissed socially or written off as third-world victims, among other things, should not ruin my day, but I'm going to go out on a limb here and say that it is our collective problem, too, because women distrusting and dismissing each other (while men get more romantic options and are presumed innocent and intellectually objective) cannot be a globally desirable paradigm for sisterhood. Yes, you read right: I wrote sisterhood. I continue to wear my suffragette colors.

The glass ceiling, unequal pay, maternal rights, war rape, genital mutilation, property ownership, literacy, violent pornography, political self-determination, birth control, higher education, and suffrage—these issues have not been resolved around the globe for women and children. Not even close.

When we talk about interracial love, we women are nervously talking about the romantic love found between men and women, but what is happening to the interracial love among all women? How are we doing?

Thinking back, I'd failed to say what I'd felt in significant moments— because I was shocked, frightened, irritated, afraid of censure and exclusion, or because I was tired and didn't want to start in on people I barely knew. That noted, I wondered how many opportunities we'd lost, how many alliances were not forged, because we had failed to really see each other, to ask the right questions, and talk through the ugly discomfort of race, sex and relationships. What is the quality of our intimacy when we confuse chatter for conversation? Girls, are we really going to fight over Charisma Man?

Seriously.

Much later, I told the book club question story to a bunch of my Asian American friends who are married to white men, and they almost passed out. They kept on saying, "No way, no way."

And I said, "Yeah. Way."

If I could have arranged time travel, I would have taken them to the book club in New York—not to have a rumble, but to ask questions and to tell our stories. I think we would have had a good cry and more laughs.

In my job as a fiction writer and as a lifelong reader of novels, I can trust that human nature will not change very much: sex will always be sold and love will always be found among strangers. It's my job to predict a character's plot and chart how things will turn out for her.

We started with a naive question about the racial politics of attraction, but our conversation taught me that the answers we had already crafted in our heads demand greater empathy and investigation. Just talking doesn't sound like much, and it is surely the lesser half of listening, but I don't see the alternative for us to get closer. In my curious days as an Asian American expatriate living in Japan, often stunned by what we say and think about each other, I wish all of us girls everywhere could really start talking about love with love.

12

THE PEDAGOGY OF THE SPIRAL
INTIMACY AND CAPTIVITY
IN A WOMEN'S PRISON

Marisa Belausteguigoitia Rius

On Friday, 1 February 2009, Tulio Lizcano, a Colombian congressman, escaped from FARC (Revolutionary Armed Forces of Colombia) guerrillas after eight years of captivity somewhere in the jungles of Colombia. During those eight years he talked to only seventeen people. He saw the sun only twice. And he engaged in an activity that saved his life. As he explains to the Mexican newspaper *Reforma*:

> I found a strategy to defeat loneliness. I cut several sticks, buried them in the ground, cut little pieces of paper, and wrote some names on them of students I used to teach in the university. I placed one name on every single stick. In the morning I would prepare class with the notebook and a pencil they had given me. During the evening, just as I had done my whole life, I taught in that imaginary classroom. I asked them questions and answered in their names. That exercise, as crazy as it sounds, kept me alive.[1]

Lizcano constructed an imaginary relationship with students and managed to address and be addressed by them. Loneliness, strict surveillance, and confinement were tamed by the creation of intimacy in a simulated atmosphere. Lizcano was whole and sane when he reached home.

This chapter focuses on pedagogical activities that foster intimacy (that is, attachment, communication, intersubjectivity) in imprisoned spaces. I focus in particular on the creation of a mural inside Santa Marta Acatitla, the most important women's prison in Mexico City. What I examine is a project that introduced color and textuality inside a prison through the design of a mural on the surface of the spiral stairs used only by visitors or by women prisoners in the process of their liberation. Two central questions emerge in relation to this project: What kind of "text" can emerge under intense surveillance? And how do captive spaces enable or resist signification?

THE BEGINNING

In October 2008 I was invited to join a project related to designing and painting the prison mural. Santa Martha Acatitla is a relatively new prison, opened in 2003. It is located in the eastern section of Mexico City, an unstable social and geographical space near one of the city's biggest garbage dumps. It is a site of relatively innovative programming, a place where prison directors are proud to offer a range of educational, recreational, and artistic activities. At the same time, Santa Martha Acatitla is far from a model institution: the rooms are small and overcrowded, the food quality is very poor, and it is dirty and punitive. Inmates there have experienced a steady erosion of both health and educational conditions. Many of the internees allude, as well, to the very conservative gender roles implicit in many of the scheduled activities—for instance, knitting, cleaning, and *maquila* (sweatshop) work. It is as if prison rehabilitation involves their "rehabilitation" to a diminished idea of women.

Vision within this detention center is tightly controlled. The prison panopticon (watch tower) allows full visibility only for the gatekeepers. Internees cannot see freely: there are walls and bars everywhere. Some of the more progressive cultural programming nonetheless has proven disruptive, and after a workshop on photography, a group of imprisoned women were unwilling to give up the privilege of vision. They wanted to continue with the experience of creating portraits, images, and thresholds of visibility inside the prison.[2] The inmates themselves came up with a new idea involving vision and unexpected frames of narration. It was a very surprising one: "We want to paint a mural in the spiral stairs."[3]

The staircase they selected was located at the southern limit of the big patio. This patio is one of the few cherished spaces inside the prison, where incarcerated women can rest and wait for their visitors, family, friends, and lovers (figure 12.1). The staircase itself contains the most anticipated or feared moment: the visit.[4] It is a prohibited space for the internees, since it connects with the outside world;

12.1 Staircase (*caracol*)

they can access it only when they are in the process of being released from prison.

Spirals are defined by their circling curving motion around a vacuum, a motion of ascent and descent. Two very important movements happen in the space of this particular spiral: family, friends, and loved ones descend to visit them, and the prisoners ascend the stairs to be released. The spiral represents the space of freedom, the connection with happiness and love to be found outside, but also pain and worry. What sort of educational enterprise were we facing when the women spoke of their desire to paint these stairs?

I was invited to work with a group of seventy inmates on a collective visual story that would ascend and descend on the spiral of these stairs. As Lauren Berlant states:

Rethinking intimacy calls out not only for redescription but transformation of analysis of the rhetorical and material conditions that enable hegemonic fantasies to thrive in the minds and the bodies of subjects while, at the same time, attachments are developing that might redirect the different routes taken by history and biography. To rethink intimacy is to appraise how we have been and how we live and how we might imagine lives that make more sense than the ones so many are living.[5]

Finding routes to imagine "life that makes more sense" could be a definition of what we searched for through the spiral and its curvy walls.

We planned a series of mural workshops inside a very stark classroom, with the goal of constructing a "monumental portrait," a narration that could signify, like a portable album, a walking memory that captured women's experience inside the prison:[6] their vision and voice inside the prison. This unusual construction could flesh out eccentric and silenced impressions and intervene in the creation of a collective visual story of captivity. Our aim was not only to spell out and articulate the remains and fragments of selves that lay inside, but to magnify them, to paint them on a wall, to turn them into a monumental story of women's lives inside prison.

Several questions soon became apparent. How could we construct a collective and monumental visual story that could represent the women's most fragile and minuscule zones of memory and intimacy? What kinds of transitions could take us from the "inside" to the outside regarding their stories and ideas of justice? Reading and writing inside prison are intensively regulated practices. Books in the library are scarce or unavailable, and documents like the national constitution or the manual on principles and laws that regulate the prison are prohibited.[7] It is very difficult to get notebooks, and paper is expensive and scarce. In this context of silencing, anything that women read and write is meaningful, especially when it is related to their intimacy or to justice.[8]

MURALISM: TO LOOK WITH ONE'S LEGS

Muralism in Mexico can be seen more generally as the monumental narration of hegemonic stories that signify heroism and patriotic love for the new postrevolutionary country. Made popular by Diego Rivera, José Clemente Orozco, and David Alfaro Siqueiros, among others, muralism multiplied after the Mexican revolution (1910–1921). In the United States there are also many important murals in schools, hospitals, and public buildings, painted by these and other artists as a means to revitalize American culture during the Great Depression. More recently, Chicano artists have appropriated muralism to depict their own history in their own way, to make sense of the nation, oppression, and meaning of being a Chicano/Chicana inside the United States.[9] This expansive use of murals reflects the fact that they are one of the best forms of both art and education for the masses. And although the contexts and meanings of murals vary, two functions endure: educating and giving pleasure.

When the great muralist Siqueiros was asked about the way in which one should look at murals, with what knowledge and with what gaze, he answered: "Murals need to be looked at with one's legs."[10] If we understand murals as a journey in signification, as ambulant meaning, or as stories "in transit," what kind of possibilities does this "motion" offer to represent intimacy within the context of the prison, where there exist so many prohibitions? The process that translated inner images, thoughts, and ideas around justice took us "inside" in two different ways: inside the prison and inside women's experiences and emotions.[11]

THE PEDAGOGY OF THE SPIRAL: INTERRELATION OF THEORIES, CONCEPTS, AND POLITICS

When deciding which feminist concepts and methods would travel effectively inside the prison and into women's intimate stories, to enable them to recount and narrate from their bodies and spaces and to visually reflect on their stories inside and "outside" the prison, three articles came quickly to mind: Gail Rubin's "Thinking Sex," Gloria Anzaldúa's "La Prieta," and Joan Scott's "Gender: A Useful Category for Historical Analysis."[12] I introduced some of the main ideas from these articles to the women in the workshop organized to figure out how "to attack the wall."[13]

This triangle of essays disrupts hegemonic notions of sexual identity and essentialist notions of seeing and knowing, as well as assumptions about experience as transparent and unquestioned. The essays convey particular notions of social justice, explore the instability of subjectivity, and provide strong critiques of heteronormativity, whiteness as the paradigmatic authority, and the essentialization of experience. "La Prieta" can be read as a diary, as a testimony of the importance of articulating sexuality and race, and of exposing the dynamics of exclusion and the obstacles for sharing what is happening inside oneself. Intimacy may be read as a path to consciousness, to the articulation of memories, to experiencing a sense of self that is necessarily fragmented and contradictory. The articles provided a three-way axis of critique against heteronormativity, whiteness, and experience as transparent; this axis was vital to creating a frame of intelligibility in which the women in prison could unravel their stories. Most of them are *prietas* (women of color), as Gloria Anzaldúa would say. Many of them identify themselves as "queer"[14]; although they do not label themselves sexually, their stories point to the practice of an unstable and undefined form of love, which indirectly points to a feminist retheorization of control and coercion in the construction of a heterosexual body.[15]

These articles and their visions helped us to identify the queerness of the wall we were about to paint. We wanted to emphasize that the construction of meaning is always inserted in some sort of captivity, that any intimate content, in order to be expressed, demands the twisting, the contortion, the curling, and rolling of language to be immersed in the grammar of signification. A wall twisted and enrolled like a spiral is the best paper, the best material to imprint and entwine dislocated and eccentric visual stories emerging from inside. The staircase as a material object curves and bends, creating the contortions needed for expression in captivity. In such a way the staircase constituted a kind of Moebius band, in which inside and outside meet and collide. Such a border allows for representation of what may emerge only at the limit of speech, signification, and language. The spiral as an oblique form and as a nonlinear process of ascent and descent, a twisted and contorted wall, could be an image of a "queer" space. The classroom we created thus offered a queer pedagogy. Its queerness may be an effect of the particular space we wanted to "attack" with form and color, and the notions we had to develop to create a nonlinear narrative to match both the spiral and the profound and oblique memories inside. The curved nature of our project, of our "twisted" classroom, called for a methodology that resembled the oblique lines and curved contours of the spiral as an inaccessible space. We, the working group, had to come up with arguments to convince authorities to allow women to occupy that space while painting it.

We also needed to look beyond these three American and/or Chicana theorists, because in the Mexican context the problematization of the concept of women has emanated from beyond queer academia or Western feminism, or lesbian theorization. Equally important for any kind of liberation are Latin-American rebellions and social movements, especially the ones enacted by the *prietos* (that is, nonwhites, the same color as the women imprisoned, who are mostly brown). These movements are very critical of the Mexican State, and especially of its relation to the democratic notion of "justice for all."

These social movements were particularly apt for our purposes because a number have incorporated murals as a pedagogical and political tool. One of the most visible uses of murals to narrate and spread out their "enclosed" history occurs within the Zapatistas movement. They have organized a rebellion that is still educating the rest of the nation, and they are using muralism as a pedagogic and political surface to illustrate what is going on in their autonomous territories, besieged by the military, in Chiapas. In symbolic synchronicity with their enclosed territories, we have named our mural *Caracoles* ("Spirals"), the same name the Zapatistas gave to their territories in August 2003.

To further grasp the political dimension of Santa Marta Acatitla's mural, it is necessary to understand the term "spirals" (*caracoles*), inspired by the Zapatista and Mayan aesthetic and political worlds. The spiral has a deep meaning in indigenous culture, especially in the Mayan one. Spirals for Mayan culture represent the reordering of the world. Through centripetal and centrifugal movements, the motion of the universe functions in ascending and descending patterns.[16] The ancient Mayan spiral and its property of condensing and producing new orders of signification have been resignified by the Zapatistas.[17] In 2003 they transformed their territory into five autonomous regions that they renamed as *caracoles*. Inside them they govern themselves through the Juntas de Buen Gobierno ("Councils for Good Governance").[18] These regions are besieged and constantly harassed by paramilitary, military, and other indigenous and mestizo groups acting in compliance with the local government. In a sense, *los caracoles* resemble an imprisoned space, due to the high surveillance and constricted movement experienced in those territories.

These five autonomous regions, these "spirals," are saturated with murals—today more than eight thousand murals.[19] These murals preserve the traditional functions of educating and providing pleasure, but one of their other aspects is really revolutionary: the place that women occupy inside this visual narration. Very soon after the uprising, on 1 January 1994, the Zapatistas had to recognize the powerful role of women inside their movement. They began to represent their leaders on the walls in their autonomous regions, and since murals are visual artifacts that provide stories that teach the relevance of historical moments or heroic personalities, they introduced women within them. In Oventik, one of the five autonomous territories constructed by the Zapatistas, there is a strong gender perspective in the design of murals: Comandante Ramona appears on the wall beside Che Guevara and one of the most important revolutionary heroes, Emiliano Zapata. Comandante Ramona was one of the female top military officers and educators who not only won a strategic battle against the forces of the state during the 1994 uprising but provided gender leadership inside and outside the Zapatista movement.[20]

The contribution of Zapatista rhetoric and discourse has reconfigured the memory, subjectivity, and intimacy of the excluded (for example, Indians, women, groups in poverty) in Mexico. The Zapatistas have created a revolution in the ways of "seeing" these others that is strongly attached to their pedagogic and artistic work on murals. Since their rebellion erupted in January 1994[21] (just when the NAFTA accords were inaugurated), the Zapatistas' re-narration of the nation in the surface of murals has been deep and unforgettable. Diverse evidence of their rebellion has been written on walls everywhere inside the Zapatistas territory, and also globally.

One event related to murals is especially important to grasp in order to understand the incorporation of the Zapatistas' narratives inside the prison. In 1997 a collective mural was painted in Taniperlas, a town located within the Lacandona jungle in the Zapatistas territory of Chiapas. The day after its inauguration, the military destroyed the mural in two different ways. First, in order to silence it, they threw white paint onto the wall. Immediately after this blanking of memory, they shot it, in order to "kill" it. Hundreds of bullets penetrated the wall. After this "massacre" of colors and stories, the Zapatistas in surrounding communities gathered together and decided to reproduce the mural of Taniperlas on multiple walls. They created "walking walls," as ambulant murals, as "stories in transit," that would tell the story of Taniperlas all over the world. The mural of Taniperlas began to be painted at sites where oppression was increasing and where political organization and commitment to the principles defended by the Zapatista movement (equality, justice, education for all, and democracy) could be found. Their story was not only to be looked at with one's legs, as Siqueiros would advise; it also had some legs of its own. The Taniperlas mural traveled to walls of those interested in reading one of the Zapatistas stories; they "attacked" the surfaces of schools, *maquiladoras* (sweatshops), and walls, not only in Mexico and the United States (such as in Tijuana and San Diego) but also in Israel and Palestine.

A special place was included in the itinerary of the *traveling wall:* prisons. Several prisons and detention centers on the northern and southern borders of Mexico were reached by this ambulant wall (as *El Amate* in Chiapas). Many migrants and political prisoners were held captive in prison at the southern and northern borders of Mexico. The borders of the Mexican nation are saturated with prisoners whose only crime is crossing them. Prisons were understood through this rebellious strategy of mural painting (as Angela Davis puts it in her book *Abolition Democracy*) as universities of revolutionaries.[22] Revolutionaries could read their history on the prison walls and look at it with their legs, in a contained space, but within incommensurable significance.

And so, along with the consideration of the notions of experience, history, and subjectivity articulated by Rubin, Anzaldúa, and Scott, we considered the ways the Zapatistas movement has changed the way we think of alterity in Mexico, especially how Zapatista indigenous women rearticulate a discussion of nation, narration, and history in highly creative, critical, and provocative ways. The Zapatistas rebellion has been exemplary for us, especially in the use of murals as nonlinear and alternative ways to narrate the other (for example, Indian, women, the poor). The Zapatistas make clear that in order to make sense you have always to move below, to descend to make known, and to narrate portions of the national history that have been buried.[23]

Needing expertise in mural design and drawing direct inspiration from the Zapatistas' spirals and murals, we contacted muralist and popular educator Gustavo Chávez Pavón (known as Guchepe), who has participated and coordinated the work on most of the murals in the Caracoles. He agreed to lend his expertise.

WORDS ON THE WALL

In order to create the mural, we needed to reflect on some of the foundations of women's memories: their past, their present, the words and images that could transmit their emotions and their experience. To this end, we organized several workshops around three notions that would represent the depth and content of their intimate stories. The first was the notion of border as the limit that separates inside and outside, freedom and captivity, below and above, surface and interior; the second idea represented justice as the failure of a promise; the third idea was the concept of emancipation as a process of liberation constructed from inside. The conceptual tool that linked all the workshops was the notion of visual autobiography.[24]

During Guchepe's (Gustavo Chávez Pavón's) first contact with the women in Santa Martha he addressed the "intimate" face of walls. He said:

> We want to transform these ugly walls for life and freedom. Art is like therapy, a way of creating intimacy between strangers, even enemies, with whom you can share your emotions. If you can't express them, your anger grows and you end up depressed. If we can transform the walls, we can transform reality. With mural painting we develop collective consciousness through personal expression of intimate thoughts and images. Murals happen like this: I come to the wall and it tells me what to paint. If you can transform a wall, you can transform your time in jail.

When asked if it was necessary to know how to paint, Guchepe answered that the idea of not knowing how to paint is a stereotype: "Of course we need a balance in the design and sometimes help to figure out an idea in visual terms, but anybody can paint. We are composers of our own emotions, and we want designs to be real, as real as the murals painted in the Zapatistas' captive zones." The goal was to leave traces of what it meant to cross the border from the exterior to the interior and the reverse, to be failed by justice, and to try to free ourselves from injustice and mental captivity by painting.

One of the most radical interventions came when we were negotiating with the director of the detention center to get official permission for women to step into this inaccessible space—the stairs—in order to paint. The spiral stairs were accessible only to women prisoners in the process of liberation or to visitors from the outside. The director established on our first approach that women could only paint "one by one." This meant that the stair could not be "appropriated" by the collectivity of women, and the process would resemble more a single silent intervention instead of a collective and scandalous one. The director granted the group access to the spiral only when we convinced her that painting a mural in collectivity in such a repressed space would convey a message of concern and interest, not only in women who were being kept locked up, but in finding and expressing artistically what women have "inside."

Gustavo Pavón explained the kind of method that would be necessary to allow the emergence of the collective visual story that could ascend and descend in spirals:

> We cannot paint one by one since our pedagogy is one of contagion. We paint with the other, within a collectivity who share thoughts, designs and colors. The only way to bring out what these women have inside is through a party, an explosion of contact and joy by the presence of colors, emotions and images. Our design, our mural, may only emerge from inside a collectivity.

The pedagogy of the spiral, the possibility of signifying enclosed experiences, can be constructed only by contagion, by silence immersed in scandalous scenarios, by infection of intimacy. Murals help us to coincide and reconcile. As our muralist said: "We may pump up balloons and free them with our thoughts written on them, like tiny surfaces that elevate and free our thoughts."

I was deeply touched and "elevated" by his words, but I rapidly landed again in our constrained reality when one of the most articulate and smart women in the workshop said: "Yes, of course, the only thing is that balloons are prohibited inside." Comments like that made me realize constantly the sort of walls we were painting on and within, and the importance of both defying such barriers and descending into the world of women.

Gustavo Chávez drew inspiration from and has himself painted murals around the world. In Palestine, for example, inside the wall that defines the occupied territories, he and his colleagues painted very quickly. As he put it, the will to stay alive made them work rapidly. They wore a mask in Palestine as a symbol of resistance and rebellion. Guchepe spoke of taking the Zapatistas' mask to Palestine as a symbol of liberation, of a process of constructing autonomy based on the pure spirit of rebellion and of acting without permission. There they painted hummingbirds and upside-down eagles (the eagle is the symbol of the Mexican nation) as signs of freedom and as symbols of being in a nation upside-down. There is always a way to leave traces.

DESCENT

After our workshop, as we were about to begin to paint, I condensed one of the most transversal and common origins of the inmates' presence as women in prison: men. During the search for the construction of the visual narrative, I heard many stories, narrations, and accounts of a triangle of betrayals: men, justice, and family. Men were the principal actors in the crimes that the female inmates had committed. Some of the women were in jail without being directly related to the crime at all; others were indirectly involved, mainly in minor acts of drug trafficking. Others did commit a crime but not the one they were charged with, and most were simply waiting to be sentenced.

Female inmates are much less visited than men; family members do not want to face the reality and the "shame" of women in prison. If the inmates have children, family members tend not to bring them to visit their mothers. The state's provision of "justice" is the biggest wound, the deepest failure, the greatest betrayer. Women's narratives had a common point of juncture or departure: the way in which men and justice exercised power over them, and the way women submitted. The mural was a space where they could rework and act upon this set of contradictions, where they could examine deeper motivations for their conduct.

The set of workshops was organized around the notions of ascent and descent, interior and exterior, of the spiral. We worked for ten sessions. Through our work on the images and the text, I began to gather information about portions of the history of the women. I learned, for example, that Cristal was reported to the police by her husband (she was a drug addict and used to have phone sex with strangers). She is a very beautiful woman, extremely smart, provocative, and rich. "I am not a saint. I took a lot of drugs, had sex phone, and risked my life many

12.2 Justice

times, but I did not commit the crime they are accusing me of. My husband told the police I participated in a kidnapping. There is no evidence at all." Cristal developed an image in her section of the mural that concentrates on her desire for sexual arousal, her resistance to authority, and her anger and desire for men. She worked on a figure that represents the dimension of justice in her experience (figure 12.2). She came up with the idea of revisiting Greek mythology and composing Justice as a beautifully classic woman, lost, blind, and overtaken by the powers of money, sex, and seduction. Her design is similar to the one Orozco painted in the building of our Supreme Court in the center of Mexico City.

CENTRIPETAL AND CENTRIFUGAL MOVEMENTS: CONSTRUCTING THE MURAL

Our spiral stair is composed of 115 square cement portions of one square meter each. Between 120 and 140 inmates participated in the mural design. Women in Santa Martha organized around the painting of the stairs and "took over" portions of its surface. Sometimes more than one woman gathered to negotiate an image within a cement square. One of the first examples of collective painting was the section done by indigenous women. Indian women are mostly in jail due to failed juridical processes. Most of the so-called legal process is illegal because of the state's failure to provide translation services.[25] Three indigenous women gathered to paint their "take" on justice. One of them, Lupita, has been sentenced to sixty years for a crime committed by her brother, who was visiting her family when the police captured him.[26] The women painted a flower, a sunflower, but with eyes that cried sixty tears, one for every year of Lupita's sentence, sixty ways to see the world as a flower, a sunflower. Doña Lupita added hands to the frame to represent the support she has been given inside prison by her inmate friends (figure 12.3).

The spiral stairs had several surfaces: the exterior one; the interior one; the base, shaped as the log of a wide tree (a tropical Mayan tree called the *ceiba*); and the summit, a huge circle facing the sky. The interior section of the spiral functioned as a Moebius band along the surface of which two very interesting images were painted. The first one was a dark fluid that streamed down the stairs to the level of the ground, a thin river of darkness, which mixed dirt and sorrow, as if the pain of the images painted above had melted inside.

12.3 Sunflowers (*girasoles*)

12.4 River of darkness

The second image emphasized the notion of time inside space. As one begins to ascend through the spiral, it is possible to see on the walls pre-Hispanic symbols representing time periods in the Nahuatl cosmovision (figure 12.4). Time measured in years ascends inside the stairs.

ASCENSION: JUSTICE

Emancipation through the oblique work on the limit of the inside and the outside has taken diverse and multiplying forms. More than ten women involved in the spiral have "walked free"; two women have begun to study law; one sent her design to a contest and won second place; others have limited family pressures and demands.[27]

We finished the mural in September 2009. Today we are expanding the spiral to other sections of the center. The news is spreading inside the prison that women who paint "walk free." Some walk free to the outside; others engage in a journey to the center of their very being, a walk inward.

The tendency of the spiral to reorder signification seems to grow and incorporate other spaces inside and outside prison. Today we are preparing another mural inside Santa Martha. I want to conclude with one image that proves the outreaching power of the spiral.

One of the most intense discussions inside our workshop was about borders. We came to the conclusion that a story can be developed only by making visible the trespassing of limits and borders: a crossing.[28] The mural text, our visual narration, performs an effect of crossing. The mural had to address a number of borders, of limits, trespassed by women: from being single to getting married, from textual to visual, the border of maternity, the line crossed when you commit a crime, from intimacy to monumentality, or when you allow yourself passively to be a victim. But the border that created the most emotion was the one that separates the United States from Mexico. Many women had family members in the United States or had lived there themselves. One of the women had a brother in the U.S. Army in Iraq, based in Samarra, and she told us that a similar spiral was created there hundreds of years ago. It is a minaret constructed during the ninth century, a huge ascending construction created to call for prayers, a spiral that needs to be looked at with one's legs. The same women told us, "When you go to the United States, let them know that we are painting this spiral of liberty."[29] We also discussed what was happening in Abu Ghraib prison and considered the power of mural painting as a way to gain consciousness.[30]

The act of gathering photos to document our work made me understand the logic of images inside prison. Women in prison are very cautious about being photographed. During the workshops the fear and distrust of being photographed was displaced, and instead they gave in to the visual urgency to depict what hurts: to kneel down before the borders we have to cross, to represent pain and hope, to construct a visual notion of justice.[31] I was eager to move them to recognize what happens when you witness your own experience transformed into visual terms. The Abu Ghraib photographic recording of atrocities constituted an extraordinary pedagogical moment to help them "see" what images "do": help to perceive the event, the experience, as really happening.

Reactions to the Abu Ghraib photographs change when viewed from within a prison. There is some understanding of the one who took some of the photographs that anatomize shame inside the prison. To look inside Sabrina Harman's emotions, to recognize her need to visually register what was happening in order to believe it, reveals the role of intimacy as a window and opportunity to develop another story.

The knowledge of a deeper and more complex story behind the images became visible when we learned about the testimony of the female soldier who recorded acts of violence. Sabrina Harman received a court notice to open her letters to the public, letters in which she stated to her girlfriend what was going on in Abu Ghraib. The letters showed that recording was a way for her to believe what she was seeing. To look through the lens and produce a photograph helped in this case, not to misrepresent through the allure of photography, but to make it real.

What sort of visual stories could the internees in Abu Ghraib or in other American prisons such as Guantanamo narrate, if we were to provide an oblique classroom and some walls? Watching the documentary *Standard Operating Procedure*, we could hear soldiers, generals, and officers, but not the prisoners.[32] The walls of Abu Ghraib remain gray and empty. They have to be painted and filled with the stories that need to be told to initiate the ascendant and descendent move of the spiral, which creates the foundations of a new narration of what America is or may become.[33]

I have written about the experiences of women and their spirals in one prison in Mexico City to let you know about their intimate emotions, desires, and concerns about justice as women, as rebels, as the weakest link of men, but also as partial owners of their stories and their lives. Women's spirals from Santa Martha reach out in curving motions to Samarra, or to any other end of the world where there is silence that could speak, as long as anybody could imagine an impossible pedagogical scene, be it a workshop where imprisoned women can access the inaccessible and represent the unrepresentable or a jungle where an

12.5 Mural

imprisoned man keeps sane by imaging a classroom—anywhere we may twist walls and create spiral streams of color in ascending and descending patterns (figure 12.5), motions of intimacy that may save our lives.

NOTES

I thank Claudia de Anda and Antonio Cintora for the invitation to paint a mural inside Santa Marta Acatitla, a female prison in Mexico City. Without them this project would not have been possible. I especially thank Gerardo Mejía for his collaboration, not only during the project but later in capturing and organizing images and preparing the text. Special thanks also to Arelhí Galicia, who managed to organize access and obtain permissions inside the center, both unimaginable activities when we witnessed the surveillance inside the prison. Irais García and Mariana Gómez organized access and tools as well as the interviews we conducted with the internees after we completed the mural. Though I was the coordinator for the event, most of the decisions around the project were taken as a group. When I use "we" I mean the participation of this group.

1. See *Reforma*, 15 February 2009 (author's translation).
2. To learn more about women in Santa Martha, see Humberto Padgett, *Historias Mexicanas de mujeres asesinas*.
3. Translator's note: *Escalera de caracol* may be translated as "spiral stairs" or "spiral staircase."
4. Women in prison are abandoned by their spouses, lovers, and family in a much higher percentage than are men. Thus the visit may be an extraordinary and very much anticipated event.
5. Laurent Berlant, *Intimacy*, 6.
6. A section of the methodology to create the mural, the one related to gender pedagogy, has been developed by Patricia Piñones, a professor of Women Studies at UNAM (National Autonomous University of Mexico) working at the Women Studies Program that I chair.
7. One of the interns asked us for a constitution. She was studying her rights and readdressing her case. To provide this, we stuck an image of a comic book (*Bird Women*) on the cover and turned the national constitution into a comic book. With such a cover, the constitution could pass into the prison and the hands of this woman.
8. More than 70 percent of the female interns are imprisoned for insignificant crimes involving thefts of less than $100 U.S. This is due to a juridical reform made in Mexico City in 2002 that increased the price of bail and made it impossible for critically poor women to make bail, even for a petty crime such as stealing a fish in a market. Only a small percentage of inmates are imprisoned because of a real crime. More than half of the crimes processed are related to a man in the family (spouse, boyfriend, father, uncle, brother,

brother-in-law). The women imprisoned by the crimes of their men are called *pagadoras* (payers). This means they are doing time instead or because of their men, since they are regularly involved with marginal or minor sections of the crime they are accused of committing. See Elena Azaola, "De mal en peor."

9. See Heather Becker, *Art for the People*.

10. Ibid.

11. We found, though, that in the process of developing the mural, authorities were either lazy about interpreting our criticism of the wall or they allowed portions of truth to be obliquely written. Nobody controlled what we painted on the wall.

12. Gayle Rubin, "Thinking Sex"; Gloria Anzaldúa, "La prieta"; Joan Scott, "Gender." The essays are quite theoretical, with the exception of Anzaldua's "La prieta," (which is easy reading and could be used as an example of autobiography). On "Thinking Sex," I encouraged the women to think of what it meant to be sexually controlled by the state (their sexuality is obsessively surveilled). Scott's essay enabled them to focus on their intimate history located at the border of reality and fiction. In other words, it offered the possibility to conceive the "truth" of their visual designs as an effect of narration and fiction, not necessarily of "real" factors. These essays also address race, sexuality, and gender as key for transformation; women in prison are also enclosed within these axes of identity.

13. "Attack" was the verb used by one of the internees, since walls are the objects that keep them away from the outside; to paint them was to rebel strongly against their function of containment.

14. In Mexico the notion of "queerness" or "queer" has traveled unevenly. Intellectuals like Norma Mogrovejo have worked with it, critiquing its dubious use for politics. (See Mogrovejo, *Un amor que se atrevió a decir su nombre*.) We use it here, as Butler and others point out, considering the oblique possibilities related to desire, writing, and subjectivity. What I want to stress here is the necessary "twist," the nonlinear methodology of such a notion, the spiral effect when we deal with sexuality in any space, but especially inside prison.

15. Sex between women in prison is common, since the normativity for sexual intimacy with men is very strict and frankly misogynist. Women in prison frequently are abandoned by their families. To be able to enjoy an intimate visit, they have to prove that they are in a stable relationship with a visitor. This is very hard to demonstrate since they do not have the possibility—while imprisoned—to engage in new relations. In short, what we face is the manufacture of celibacy and the proliferation, in some cases forced, of queer and lesbian sexual relations inside prison. Notwithstanding this qualification, some women report that the best sex and the most profound love they have experienced has been with women in prison. See Marisa Belausteguigoitia, "Mujeres en espiral."

16. See Miguel León-Portilla, *Los antiguos Mexicanos a través de sus crónicas y cantares*.

17. The Zapatistas rebellion erupted on 1 January 1994. From 1994 to 1996, the rebels negotiated with the government precise changes in the constitution in order to be granted cultural and political autonomy. In 2001 Congress voted against this "contract." Today the

Zapatistas do not accept economic support from the government and govern their regions autonomously with strong surveillance and intervention by the military, paramilitary, and disguised government actors, hidden behind NGOs like OPDDIC (Organización para la Defensa de los Derechos Indígenas y Campesinos A.C. / Organization for the Defense of Indigenous and Peasants' Rights).

18. The five designated regions known as *caracoles* (spirals) are: La Realidad, Oventic, La Garrucha, Roberto Barrios, and Morelia. Today they are particularly harassed; Morelia has been attacked by official organizations such as OPDDIC.

19. For more on *muralitos zapatistas*, see Luis Adrian Vargas, *El muralismo Zapatista en Oventic*.

20. See Márgara Millán, "Las Zapatistas del fin del milenio."

21. In Mexico, for decades murals have been a paradigmatic resource to narrate monumental episodes of our national history. The Zapatista movement has constantly reappropriated segments of the nation's history through the construction of alternative narrations that have made visible the unrecorded experiences of Indians and, today, of women. To see more on the uses of murals inside Mexican history, see Vargas, *El muralismo*.

22. See Angela Davis, *Abolition Democracy*, 7.

23. At the beginning of the rebellion, Subcomandante Marcos communicated with a distinctive strategy. He attached postscripts to his communiqués in order to insert different levels of communication from below. Soon the postscripts began to multiply to the point of creating a wide list in one communiqué. For more on these strategies of descent, see Marisa Belausteguigoitia, "Máscaras y postdatas."

24. For more on the notion of visual autobiography, see Anna Marie Guasch, *Autobiografías visuales*.

25. In Mexico we find more than 60 ethnic languages; more than 50 percent of indigenous women do not speak Spanish.

26. Commenting on the types of offenses committed by indigenous women is very complex. Access to their cases is very difficult and granted only to their lawyers. We rely on the stories the prisoners have told us. Their own accounts, complemented by the literature around the perverse justice system in Mexico, establish that most offenses were related to men dealing with drugs. Women are the "weakest link," often simply wrongly accused of crimes due to failed translations; most Indian women do not speak Spanish fluently, and although translation is legally required it frequently is not provided by the state (see Azaola, "De mal en peor").

27. We have not yet done the necessary research to know the reason for this number of women "walking free." Of course, painting is not one of them, but the image of them ascending to freedom was a very powerful one for the conception of the phrase "the one who paints walks out."

28. According to Ricardo Piglia, a story is the result of the crossing of a special border, the one that lies between a secret and its unveiling: Ricardo Piglia, *Formas breves*.

29. Gloria Anzaldúa gives an account of the complexity of stories emanating from the border between the United States and Mexico, where an enormous wall is being both erected and appropriated by painting: Gloria Anzaldúa, *Borderlands/La Frontera*).

30. Mark Padilla et al. warn us, however, that local accounts of intimacy should not be blind to the ways in which globalization permeates these affective and social relations: Mark Padilla et al., *Love and Globalization*.

31. Virginia Woolf, in *Three Guineas*, and later Susan Sontag, in *Regarding the Pain of Others*, explore the powerful effect that images may have on the cause "to end the war." Both writers analyze the power of photographs—of images—to reveal and mislead.

32. *Standard Operating Procedure* is a 2008 documentary film, directed by Errol Morris, that explores the meaning of the photographs taken by the police and army at the Abu Ghraib prison.

33. In Madrid, popular educators conducted a project whose goal was to paint the walls of Carabanchel, a prison for Spanish Republicans (leftists) during the Spanish Civil War and afterward. One night in November 2007, police came with bulldozers and destroyed Carabanchel. The practice of painting the walls of prisons is becoming more and more popular. See http://salvemoscarabanchel.blogspot.com/.

BIBLIOGRAPHY

Anzaldúa, Gloria. "La prieta." In Cherríe Moraga and Ana Castillo, eds., *This Bridge Called My Back*. San Francisco: ISM Press Books, 1988.

——. *Borderlands/La Frontera. The New Mestiza*. San Francisco: Aunt Lute Books, 1999.

Azaola, Elena. "De mal en peor: Las condiciones de vida en las cárceles Mexicanas." *Nueva Sociedad*. Mexico: CIESAS, Plaza y Valdés, 2006.

Becker, Heather. *Art for the People: The Rediscovery and Preservation of Progressive- and WPA-Era Murals in the Chicago Public Schools, 1904–1943*. San Francisco: Chronicle Books, 2002.

Belausteguigoitia, Marisa. "Máscaras y postdatas: Estrategias femeninas en la rebelión indígena de Chiapas." *Debate Feminista,* año 6, vol. 12. Mexico, 1995.

——. "Mujeres en Espiral: Justicia y cultura en espacios de reclusión." In *Experiencias en territorio: Género y gestión cultural*. Mexico: PUEG, UNIFEM (forthcoming).

Berlant, Lauren. *Intimacy*. Chicago: University of Chicago Press, 2000.

Butler, Judith, and Joan Scott. *Feminists Theorize the Political*. New York: Routledge, 1992.

Davis, Angela. *Abolition Democracy: Beyond Prison, Torture, and Empire*. New York: Seven Stories Press, 2005.

El libro de los libros del Chilam Balam. Mexico: Fondo de Cultura Económica, 2004.

EZLN. Documentos y Comunicados. Mexico: ERA, 1994.

Guasch, Anna Maria. *Autobiografías visuales*. Mexico: Siruela, 2009.

León-Portilla, Miguel. *Los antiguos Mexicanos a través de sus crónicas y cantares*. Mexico: Fondo de Cultura Económica, 2005.

Millán, Márgara. "Las Zapatistas del fin del milenio. Hacia políticas de autorepresentación de las mujeres indígenas." In Belausteguigoitia Marisa and Leñero Martha, eds., *Fronteras y cruces: Cartografía de escenarios culturales Latinoamericanos*. Mexico: PUEG/UNAM, 2005.

Mogrovejo, Norma. *Un amor que se atrevió a decir su nombre: La lucha de las lesbianas y sus relaciones con los movimientos homosexuales y feminista en América Latina.* Mexico: Centro de Documentación y Archivo Histórico Lésbico (CDAHL), 2000.

Padgett, Humberto. *Historias Mexicanas de mujeres asesinas.* Mexico: Planeta, 2008.

Padilla, Mark, Jennifer Hirsch, Miguel Muñoz-Laboy, Robert Sember, and Richard Parquer. *Love and Globalization: Transformations of Intimacy in the Contemporary World.* Nashville: Vanderbilt University Press, 2007.

Piglia, Ricardo. *Formas breves.* Buenos Aires: Temas, 1999.

Rubin, Gayle. "Thinking Sex: Notes for a Radical Theory of the Politics of Sexuality." In Carole Vance, ed., *Pleasure and Danger.* London: Pandora, 1989.

Scott, Joan. "Gender: A Useful Category for Historical Analysis." *American Historical Review* 91, no. 5 (December 1986): 1053–75. For the article in Spanish, see Joan Scott, "El Género como una categoría útil para el análisis histórico." In Marta Lamas, ed., *La construcción de la categoría de género.* México: PUEG/UNAM, 1996.

Sontag, Susan. *Regarding the Pain of Others.* New York: Picador, 2004.

Vargas, Luis Adrian. *El muralismo Zapatista en Oventic, Chiapas.* Mexico: UNAM, 2007.

Woolf, Virginia. *Three Guineas.* San Diego: Harcourt Brace Jovanovich, 1966.

IV

GLOBAL
FEMINISM
AND
THE
SUBJECTS
OF
KNOWLEDGE

13

WITNESSING, FEMICIDE, AND A POLITICS OF THE FAMILIAR

Melissa W. Wright

> *We have much work left to do, the road ahead is long and hard. There will come a time when my voice becomes silent so that new voices can be heard to carry on the struggle for the rights of women, which, as I have said, is also for the rights of men, because it is the struggle for a more just and democratic society for all.*
>
> ESTHER CHÁVEZ CANO, 9 NOVEMBER 2007, IN MOLLY MOLLOY,
> "THE WOMAN WHO DARED TO STAND TALL ON THE BORDER"

> *At what cost do I establish the familiar as the criterion by which a human life is grievable?*
>
> JUDITH BUTLER, *PRECARIOUS LIFE*

What is the cost of establishing familiarity as the criterion for mobilizing political action? This question, which is at the heart of Judith Butler's query above, is central to discussions within the human rights literature regarding the political advantages and disadvantages of forming social justice movements around the politics of testimonial witnessing, a strategy that hinges upon establishing familiarity between the testifier and her public as a way to create a political community of witnesses to injustice.[1]

In this essay, I examine this question in relation to a social movement in northern Mexico that, in the mid-1990s, galvanized political action against the killing of women within a climate of state-sanctioned impunity. This movement began in the border city Ciudad Juárez, where activists began documenting the murders and kidnappings of several hundred women and girls; they brought attention to the lack of political will either to stop the murders or investigate the crimes. By the end of the decade, the activists called this violence femicide, and they succeeded in laying the groundwork for an international campaign against femicide not just in Ciudad Juárez but also in other parts of the world.[2] In focusing on the violence against the women and girls of Ciudad Juárez, the activists generated a broader critique of corruption in the Mexican judicial and political system that created the conditions of impunity enjoyed by the criminals, and they criticized the export-processing factories (*"maquiladoras" or "maquilas"*) for the poverty wages and inadequate transportation systems that exploited the vulnerability of the working women whose poverty made them even more vulnerable to the criminals who preyed upon them. And the movement also brought attention to the harassment of activists and of family members of the victims who demanded competent police work and accountable governance on the part of elected officials.

While the movement did not fulfill all of its aims, and while competent investigations of the crimes remain elusive, the anti-femicide campaign politicized violence against women along Mexico's northern border and showed it to be not just a matter of public safety and gender but also, fundamentally, a matter of democracy and accountable government. An important legal victory occurred on the international stage in 2009 when the Inter-American Commission of Human Rights sanctioned the Mexican government for failing to protect the life and integrity of women in Ciudad Juárez. Domestically, however, such hard-fought victories are on the threshold of being lost as the social movement that brought international attention to the femicides has splintered over internal disputes and has been unable to galvanize against the overwhelming violence that currently afflicts Ciudad Juárez. Since 2007, more than 50,000 people have died violently in the country; about 20 percent of those have perished in Ciudad Juárez, along the northern border, making this city one of the most violent in the world. The violence is commonly referred to as "drug-violence," but what is causing it, who the criminals and victims are and the factors contributing to its spiraling, are still matters of conjecture in the absence of sound investigations and professional policing. What is known is that the victims come from all corners of life, and among the victims are the ongoing signs of femicide, of women being killed for being women and in a context of impunity.[3]

In relation to this turn of events, I examine how the femicide movement's development around a politics of testimonial witnessing contributed to its accomplishments as well as to its weakening in recent years, especially in the context of the unprecedented violence that is overwhelming Ciudad Juárez. I focus on the issue of witnessing as a means for examining the internal strategies of the femicide movement as it responded, at different moments, to the need to establish familiarity as a criterion for mobilizing political connections across geographic scales. As my discussion will illustrate, "familiarity" is a fungible concept, its shape and content shifting along with the production of testifying subjects and their audience, as they create or fail to create a political connection around the ongoing chain of witnessing. For as Jacques Derrida insightfully observed in *Force of Law*, there are no original witnesses. Witnessing is, rather, a dialectical cycle of address and response around the impossibility of there ever being an original witness. Witnesses bring about witnesses.[4] For this reason, as Nora Strejelivich, a writer, activist, and torture survivor from Argentina, has written, the politics of witnessing requires that the testifying witnesses meet "the expectations readers or listeners have regarding what truth means and how it should be voiced."[5] In short, testifying witnesses must offer a testimony that appears, and makes them appear, familiar as a certain kind of political speech that, in turn, politicizes their witnessing public.[6]

The femicide movement's attempts to meet this expectation for familiarity exacted some political costs, especially for its networks and efficacy within Mexico as a force for motivating pressure on elected officials. I examine such attempts by focusing on how the movement responded to attacks on the victims and on the activists for being "women in the street," or what I call "public women."[7] Such attacks have persistently portrayed both the victims and activists as "abnormal" or "perverse" women, and the activists have responded by using a strategy for presenting the victims and activists as "family women," as women familiarized to their public through their representation as "daughters" and "mothers." As I explore here, the mothers' use of testimonial speech, as "testifying mothers," has been the principal means by which they have established not only these identities but also an intimate connection with their witnessing public that witnesses the pain of mothers for their daughters. In making my argument, I draw from an extensive literature, much of it feminist and from multiple disciplines, that critically examines the representations of motherhood, as a particularly situated social and political identity, in the making of testimonial speech and the testimonial subject.[8] This scholarship lays much of the foundation for my exploration into how the intimate experience of witnessing a mother's visceral pain and grief creates political bonds that connect the intimate to the global, as mothers' tears motivate public outrage over injustice.

This discussion entails following the movement from its inception in Ciudad Juárez in the 1990s as a collaborative effort among feminists, community activists, and family members of femicide victims. By tracking how the movement established the criterion of familiarity in the first years of its activity, during which it made notable gains in the domestic context, I lay the groundwork for understanding the political costs as the femicide movement had to respond to shifting expectations among its witnessing public for what constituted familiarity as the key criterion for creating a political community across domestic and international social justice networks. And finally I conclude with some speculative remarks regarding how the need to produce familiarity as the bond binding testifying witnesses to their witnessing public is currently limiting the femicide activists' ability to generate public protest over the escalation of violence and the curtailment of citizens' rights in relation to the government's drug war.

FROM DEAD GIRLS TO DEAD DAUGHTERS

When a handful of women occupied the Ciudad Juárez mayor's office in 1993 to protest the lack of government action against the violence, they declared to the press that they were witnesses to the crimes. "They thought no one was paying attention," said Esther Chávez Cano, one of the original instigators of the femicide movement. "When the press came, that's when we told them that we had a list of *las muertas* (the dead women/girls)." And with that list, and that occupation, the protests over "Las Muertas de Ciudad Juárez" (The dead women/girls of Ciudad Juárez) began.[9]

The women in this original occupation had been active in various causes, such as in feminist organizing around reproductive rights and in the fight to make the government accountable for the extrajudicial kidnappings and murders that occurred during the "Dirty War" of the 1960s and 1970s. They had initially headed to the mayor's office with the intention of meeting with him, but when they realized that the mayor had stepped out of his office and had stood them up, they decided to occupy his office. "And we came up with our name on the spot," said Esther Chávez Cano, who had compiled the list from reports she had culled from the back pages of the city's newspapers. "We called ourselves 'El 8 de Marzo de Ciudad Juárez,'" after the feminist organization based in Chihuahua that was leading the fight to make domestic violence a crime in the state.[10]

Within a year, El 8 de Marzo de Ciudad Juárez, with Chávez as its spokeswoman (*vocera*), joined with several other organizations in the city to form The

Coalition of Non-governmental Organizations for Women (La Coodinadora de ONGs en Pro de la Mujer), which I shall refer to as "the Coalition." Eventually, the Coalition would consist of some fourteen different women-led organizations, which worked on projects ranging from public health, community development, domestic violence, youth at risk, and other related causes. Family and friends of the victims participated as individuals in the Coalition, until they formed their own organization, Voces sin Eco (Voices without a Sound) in 1998.

Soon after, the Coalition created a list of demands that centered around three principle ideas: One, that the city government and the city's export-processing firms implement strategies for preventing further deaths and kidnappings; two, that the Chihuahua state conduct competent investigations into the crimes already committed; and three, that governing elites (at all levels), along with civil society, address the cultural context that justified violence against women and that created the conditions for femicide with impunity.[11] They justified their demands with the idea that they were witnesses to the government's impunity. As Esther Chávez told me some ten years later, "Nobody cared about these poor girls because they were poor. That attitude created the impunity, and we let them (governing officials) know that people in this city cared about these girls. They just had to know it was happening."[12] And so to publicize the violence, the Coalition had to bring word of it out of the newspaper back pages and put it front and center so that the city's residents, too, could witness its occurrence in their neighborhoods and city streets. With such aims in mind, the Coalition organized marches, press conferences, confrontations with public officials, and other events that, in effect, created a witnessing public for its own witnessing of the conjoined crimes of murder and impunity.

In communicating its demands, the Coalition made connections between the violence and the political economy of export-processing that had been the engine of growth along Mexico's northern border and that had laid the path for the implementation of the North American Free Trade Agreement (NAFTA) in 1994. The group called for better policing and lighting along the dark roads that women and girls had to walk as they commuted to their jobs in the city's export-processing industries (the *maquiladoras*). It urged employers and urban administrators to provide livable wages and safe transportation to the workers and to improve the living conditions of the city's working poor, the majority of whom lived in improvised housing with inadequate services.

To generate force behind these demands, the femicide protestors created national and international networks that brought attention to the violence and made Ciudad Juárez infamous not only for the feminization of the international division of labor but also for the murdering of these very working women. The city of *obreras* ("female workers") was becoming the city of *las muertas* ("the dead girls"). As a result, domestic and international political and consumer organizations pressured the political leaders of Mexico as well as the leaders of international corporations doing business in the country to stop the murders and the impunity that protected the criminals. And the government and corporate leaders responded. By 1996, officials in statewide and federal office were increasingly having to answer questions regarding their actions to stop the killing of women in Ciudad Juárez. "They were shocked," Chávez later told me. "They couldn't believe that a bunch of women could do what we did."[13]

But while the political and corporate elites responded to the public pressure created by the Coalition's strategies for creating public witnesses both to the crimes and to the impunity, they also fought to stem public support, and they did so by dismissing the victims as worthy of any public concern. They made this argument by associating the victims with "public women" (*mujeres públicas*) as a way to diminish the idea of the victims' innocence. In Mexico, the term "public woman" suggests the negative interpretation of a prostitute (*la puta*) who represents the "fallen woman" whose uncontrolled sexuality represents a contagion and a threat to society.[14] Public women, according to this widespread and familiar discourse, create the very trouble that they experience; by contrast, the term *hombre público* ("public man") is another way of saying "citizen."[15]

The political and corporate elites used this discourse as a way to weaken public sympathy for the victims of the violence and dilute the public pressure on them to prioritize women's safety. In claiming that the victims were public women who actually caused the violence that ended their lives, they referred to a line of argument that in its extreme actually justifies the violence against women as a way to rid society of trouble. If public women are the source of the violence, then that they are being killed by this violence is a means for ending its root causes. With this strategy they countered the Coalition's witnessing to impunity by presenting their own witnesses to the victims' guilt of the crimes that ended their lives. In such a context, they contended, there is no impunity, just the normal events in a city full of public women. This position was given official credence by the then-governor, Francisco Barrio, who in spite of creating the special prosecutor's office, dismissed the violence as "normal" for the city.[16] The Coalition was creating, therefore, witnesses to nothing wrong, just to the everyday occurrences in a city full of public women.

The Coalition fought back with a retooling of its witnessing strategy and declared that it was not only a witness to the impunity of the state but also to the innocence of the victims. "There was a change in what we were doing," one activist later told me, "We realized that we had to stand up for the victims too. It wasn't enough to protest against the government."[17] The protestors created this change by declaring the victims to be "innocent daughters," who had been doubly victimized by those who murdered them and by the state that failed to protect them as they worked in the factories that had fueled Mexico's industrial growth.[18] Esther Chávez explained this shift, "We had to make people realize that they were real people, with families, with people who loved them and missed them. The families already called them *hijas*, ["daughters"] and so we did too."[19] While many of the activists did not intend for this discursive shift to result in an exclusion of sex workers and domestic violence victims from their protests, some divisions began to emerge within the movement over this issue.[20]

This approach reveals the centrality of empathy to the politics of witnessing that the Coalition embraced. As the activists began referring to the victims as *las hijas* rather than as *las muertas*, the victims gained a familial identity that made them familiar against the government's accusations that the victims had forsaken their familial identities through their public activities that had turned them into "whores," or non-familial women. Such a strategy has been common to social movements, particularly during the dictatorships across Latin America, when protestors referred to victims as "children" as a way to diffuse government characterizations of them as "communists" and "terrorists," among other unsavory labels. For instance, Las Madres de la Plaza de Mayo ("Mothers of the Plaza de Mayo") employed this strategy to great effect as they succeeded in generating a social justice movement that helped bring down a violent military dictatorship. By referring to the victims of state-sanctioned kidnappings, torture, and murder as *hijos/as*, Las Madres created international empathy for the victims that was linked directly to outrage against a government that would murder innocent children.

Likewise, the Coalition in Ciudad Juárez employed such a logic as they worked to generate public empathy for the daughters who had been murdered as they walked to their factory jobs, did their shopping, and waited for their buses on the city's streets. And they successfully linked this empathy to a protest against a city and state government that did nothing to protect "las hijas de Ciudad Juárez." As one participant in the protests put it, "When people began to see the victims as daughters, the government had to do something. What kind of government doesn't do anything to protect its daughters? An illegitimate one."[21]

And by 1998 the public empathy for the victims and their identification as daughters had gained international exposure as the *New York Times*, CNN, and other international news sources, in addition to Mexico's major newspapers, criticized the Mexican government for allowing Ciudad Juárez to become a place where factory-working daughters were being stalked with murderous impunity as they commuted to their jobs.[22] In that same year, the Chihuahua state government established a special prosecutor's office to investigate the crimes, and the Mexican National Commission on Human Rights declared, for the first time, that *sexism* was a political problem that contributed to government-sponsored impunity.

As part of this success, Chávez, with the assistance of other activists, formed the city's first sexual assault and rape crisis center, Casa Amiga, and public officials and corporate leaders were on the defensive as they kept trying to generate public distrust for the victims. As the director of the Maquiladora Association in Ciudad Juárez (AMAC) stated in a televised interview on the U.S. network ABC's news program, *20/20*, in 1999: "Where were these young ladies where they were last seen? Were they partying? Were they on a dark street?"[23] The question, in other words, is what were these women and girls doing out in public?

But as public outrage over the killing of Juárez's daughters increased, along with pressure on political and corporate leaders to do something about the crimes, this attempt to declare the victims to be responsible as public women for the crimes failed to destabilize the Coalition. Thus, faced with the Coalition's acumen in connecting its witnessing for innocent daughters to a movement that witnessed impunity, the elites took aim at the activists. By 1999, political and corporate references to the victims as public women began to wane as officials set their sights on the activists.

PROFIT, PROSTITUTION, AND PUBLIC WOMEN

The initial target of the public woman discourse was the leadership of the Coalition, namely Esther Chávez, who had gained fame as the movement's "spokesperson" (*la vocera*). Her founding of the rape crisis and sexual assault center, Casa Amiga, had also demonstrated her acumen in raising funds and even political support, as she had received some financing from the city government. But her success also contributed to some of the tensions within the Coalition. Esther, who died in December 2009, was a feminist and believed strongly that feminist analysis and politics was necessary for fighting femicide. But not all of the organizations, although all women-led, were comfortable being known for "feminism," a moniker that continues to invoke "men-hating" and women who forsake their domestic duties. Esther, who had never married, who had no children, and who had retired from a successful career as an executive accountant in a multinational corporation before politically organizing for women's reproductive rights and then against femicide, was clearly, as she described herself, "a shameless feminist" (una feminista sin vergüenza). Her critique, particularly of domestic violence, and her founding of Casa Amiga as a place that attended prominently to victims of family abuse exposed fault lines within the Coalition between those who supported Esther as the spokesperson and feminism as a political approach, and those who did not, especially those who did not want to critique the family as a political entity.

Adding to the political problems were the ongoing tensions of class. The Coalition's cross-class alliances, in which middle class, wealthy, and extremely poor women collaborated in the same organization, were showing signs of strain. While some of the women who participated in the Coalition received salaries as personnel of their organizations, others did not and offered their time voluntarily. For the wealthier women of a couple of organizations, this volunteering posed no economic stresses, but for the unpaid women who came from the city's working poor, their activism required time and resources that were scarce and created hardships for themselves and their families. This class disparity came out publicly after the forming of the victims' families organization, Voces sin Eco, whose members came from the working poor. Their struggles to participate in public events, to find the resources for transportation and child care, to take time off from work, and so forth, contrasted with the experiences of other Coalition members who either participated in events as an offshoot of their professional activist activities or who had no economic worries. When Esther, a middle-class professional woman, formed Casa Amiga and gained further notoriety as an activist who would now be a paid director of her own new organization, her financial acumen, not just in founding the organization but in founding her own job, exacerbated these class tensions around the accusation, from a minority within the Coalition, that Esther was profiting from the movement for her own political and economic purposes.[24] Such attacks took on the shape of the public woman discourse; Esther was portrayed as someone who was taking advantage of the victims by selling their stories to an international public as a way to further her own economic and political goals. The accusations went further in denouncing Esther for also destroying the reputation of the city by constantly raising the issues of violence and sexual assault. So, like the worst connotation of a prostitute, Esther was defiling not only herself through her public activities, she was also defiling the families and communities of the city.

The attacks against her became so vicious that anyone associated with Esther was liable to also become a target of this public woman discourse. For the other organizations, working with Esther in the Coalition was becoming a liability that could jeopardize public support and funding for their own causes. As one member of the Coalition told me in February 2007, "We couldn't even support her. We didn't want to get burned with her." And under this pressure, the Coalition fell apart by the end of 1999; with that collapse, the social movement against femicide lost its infrastructural center. The family organization Voces sin Eco also dissolved by 2000. Meanwhile, the murders continued.

The horrible discovery of eight bodies of women and girls, all showing signs of prolonged torture, in central Ciudad Juárez in November 2001 reinvigorated the movement as the evidence of femicide was undeniable. The Coalition temporarily reformed as Esther and other organization leaders held a press conference to denounce the murders and to present witness, once again, to the ongoing femicide. Outrage over the violence sparked a gathering of organizations in the state capital of Chihuahua City, where other women-led organizations collaborated in the coordination of a protest march, called Éxodo por la Vida (Exodus for Life), across the 220 miles (360 kilometers) of desert separating Ciudad Juárez from Chihuahua City. But while these organizations were seasoned in the politics of activism across a number of causes, from reproductive rights to democratization to home-ownership rights, they demonstrated that they had learned from the public woman attack on the Coalition, and on Esther especially. When they marched across the desert, they wore the black clothing of a woman in mourning and as they placed crosses in the desert, they used religious symbols to represent the victims. Referred to by the press as Las Mujeres de Negro ("The Women in Black"), the Chihuahua group was greeted by hundreds as they entered Ciudad Juárez and marched, through the city center, to erect a cross at the downtown bridge that led into El Paso, Texas. The symbolism of the march indicated a marked shift from a previous round of organizing, as the Women in Black and the religious symbols indicated the domestic and private nature of the activists on the street. And the two family-identified organizations that formed out of this event illustrated this shift clearly as they took over the leadership roles of the movement. One, Justicia para Nuestras Hijas ("Justice for our Daughters"), was based in Chihuahua City and the other, Nuestras Hijas de Regreso a Casa ("Bring Our Daughters Home") in Ciudad Juárez. A third family organization, The Women's Institute of Chihuahua (ICHIMU), later formed, also in Ciudad Juárez, and was directly affiliated with the state government office, even as it was under the direction of one of the former leading activists of the Coalition.

With the formation of the family organizations as the new leading voices of the femicide movement, their spokespeople declared themselves to be activists whose political activities emanated from their experience as family women rather than as public women driven by overtly public concerns. And unlike the activists within the Coalition, the family organization spokespeople presented themselves as either mothers of the victims or as people speaking out on their behalf. This declaration was a frontal rebuke of the public woman accusation that had been so destructive for the Coalition and that had weakened public support for the movement within Mexico. Mother-activists thus became the central figures in the femicide movement after 2001, and they stood in stark contrast to the public women of the previous decade.

The declaration of mother-activism materialized primarily through the practice of testimonial speech that the mothers within the organizations used to demand justice for their daughters along with an end to impunity. While testimonial speech had always been a part of the femicide movement and had certainly been key for turning "las muertas de Ciudad Juárez" into "las hijas de Ciudad Juárez" as mothers testified to the daughterly innocence of the victims,[25] the use of testimonial speech as a way to witness on behalf of the activists became a defining feature of the movement in its second decade. The mother-activists used testimonial speech, in other words, not just to testify on behalf of their daughters' innocence and therefore on behalf of their validity as victims, but also on their own behalf as family women who were not, like public women, driven into the public sphere for their own selfish reasons.

In this context, testimony refers to the meaning of *testimonio*, the Spanish term for legal testimony that took on a key, and nonlegal, meaning in the late twentieth century as a human-rights strategy in response to the oppressive regimes of Latin America. The practice of *testimonio* has been especially effective for providing women from the economically impoverished classes a means for articulating their social justice demands within the structures of social movements across the region.[26] In this usage, *testimonio* is a first-person account of an injustice perpetrated or sanctioned by the state, and the testimony is offered not in a public court of law but in the public realm of opinion with the purpose of motivating this public to demand justice from governing elites. The power of *testimonio* derives from the testifier's credibility as someone whose personal experience of injury reflects a collective experience of injustice.[27] The testifier's credibility as a witness to the public nature of the private injustice thus lies in her ability to present that injustice in a way that rings familiar to her witnessing public; the testifier witnessed the injustice personally, while the public witnesses the atrocity publicly.[28]

For this reason, testimonial speech relies heavily on the presentation of the testifier's credentials as those qualifying her for the role of witness. The witnessing of Rigoberta Menchú provides a good illustration of this imperative. For example, when Menchú described her experiences in her testimonial, *I, Rigoberta Menchú*, she based the validity of her denouncement of the genocide perpetrated by the Guatemalan army upon her own experience of them; as an indigenous woman, she made the claim that her personal experience of the injustice was also the collective experience for other indigenous peoples in the country and beyond. Later, however, when the veracity of her account was called into question, the issue was not whether the events she had described had occurred—the genocide had been documented by other sources—but whether these events had happened to her personally.[29] Her firsthand experience of these events was fundamental to her claim that she could bear witness through her testimony to the genocide of indigenous peoples at the hands of the Guatemalan state, but when her firsthand experience was challenged, her ability to stand as a witness to public injustice was weakened.

The mother-activists within the northern Mexico femicide movement have shown their understanding in terms of minding this imperative for establishing their firsthand experience of personal injury as a way to make public demands, on behalf of others, for justice. Their claims rest on their declarations that they, as mothers, have lost their daughters to the violence. As such, they are the only ones with the legitimacy to testify to the injustice of a child's murder and a state's indifference to that murder. And their testimonies often illustrate the heart-wrenching experiences of this injustice as they describe their children, their experience of being their mothers, their realization of their kidnappings and/or murders, of their children's suffering, and then of the state's indifference and, often, derision as they sought help in finding the killers and doing something to stop the violence. The emotion within their testimonies is palpable, along with the expressions of anger and sadness, and their force derives from the message that their experience lies in the experience of a mother who has lost a child under the most terrible of circumstances. They are not, in other words, radical feminists who seek to overturn the patriarchal structures of family and religion. They are, rather, mothers who will do anything for their children. They will even go out onto the street, raise their voices, march, and be disruptive, as they look for their children with the hope of bringing them home. They

are doing what the most conservative elements of religious and family institutions sanction around the world: they are trying to take care of their
children.

Yet, like mother-activists in other places and times, those in the femicide
movement use this testimony in order to make extremely radical demands. They
are calling for the end of corruption from a corrupt state; they are calling for
competent police investigations from incompetent police forces; they are calling
for investments to make city streets safer for the working poor in a place where
the safety of the working poor is not a priority; and they are demanding respect
for young women workers who have endured decades of derision for fueling the
country's industrialization. But, perhaps most radically, like the Mothers in the
Plaza de Mayo in Argentina, like the Mothers of the disappeared in El Salvador
and in Guatemala, like the mothers who have marched in Chile and in Brazil,
these mothers are promising that they will return home when their children return. They make this promise even when they know their children to be dead.
For this reason, mother-activists emerge as a most radical kind of activist, a kind
of eternal revolutionary whose morally sanctioned protest will never end.[30] Yet,
these radical demands are not based on their own radical identities, but rather
upon the conservative conception of mothers as those who will sacrifice themselves for their children's welfare.[31] They live for their children. Consequently,
through their testimonies that establish their credibility as witnesses to state corruption and impunity, the mother-activists turn such conservative logic on its
head to make extremely radical demands of the political and corporate establishment. But these demands rest squarely on their constant reiteration that they
are fundamentally private women whose public presence on the street indicates
that something is terribly wrong in their communities. The underlying logic reveals a reiteration of the discourse within the public woman refrain: women on
the street mean trouble, but in the case of private women on the street, the
trouble is external to them.

Another important quality of testimonial speech that is reproduced through
the mother-activist testimony is that of its simplification of complicated political issues via a homogenization of the testimonial message.[32] For even as the
mothers' narratives vary according to their details, they share the same basic
structure that identifies the political message as pertaining to a mother-activist,
as opposed to a different kind of activist and a different kind of politics. All of
the mothers reiterate that their personal political motivations derive from their
experience as mothers who have lost daughters to femicide. As such, their political activism is an outgrowth of a family relationship that is familiar the world
over. Anyone who is familiar with mothers and with their feelings for their

children can therefore identify with them. Anyone who is familiar with the experience of being someone's child can identify with this mother's politics, derived from her love, grief, outrage, and sadness. Mother-activism therefore emerges as a most familiar kind of activism based on the most familiar kinds of activists: mothers, who are familiar because of their familial presence the world over. The differences of language, class, and regional experience are unimportant in a story of mothers looking for their children, a story that for its simplicity and familiarity seems to transcend politics and the vagaries of geography, history, and culture. Mother-activists, therefore, like their causes and the justice they seek, are both intimate and global.

And on the strength of mother-activism, the family organizations within the femicide movement have succeeded in expanding the international public for the movement. As one activist in Ciudad Juárez said, "The mothers tell the same stories because that's what the audience wants to hear, even though they have heard it before."[33] Through their testimonial speech and tireless activism, they have internationalized the movement well beyond its reach in the 1990s. By creating a political community beyond the country, the mothers have created constant pressure on the Mexican government to answer its critics and address, especially, the issue of impunity. A major victory as a result of their testimonial witnessing was handed down by the Inter-American Commission of Human Rights in December 2009.[34] In a ruling with potential implications across the Americas, the Court determined that the Mexican government violated the American Convention on Human Rights and the 1994 Inter-American Convention on the Prevention, Punishment, and Eradication of Violence against Women by failing to prevent the murders and to investigate the crimes.

The Court declared that Mexico must do the following:

In addition to conducting a serious murder investigation and investigating law enforcement officials responsible for obstructing the cotton field case, which included the fabrication of scapegoats under torture, within one year the Mexican government must hold a public ceremony in Ciudad Juárez to apologize for the crimes; build a monument to the three murdered women in the border city; publish the sentence in the official government record and in newspapers; expand gender sensitivity and human rights training for police; step up and coordinate efforts to find missing women; permanently publicize the cases of disappeared women on the Internet; and investigate reported death threats and harassment against members of the families.[35]

This victory in the international court is noteworthy for many reasons; it is the first case of femicide to be ruled on by the court; it is binding on the Mexican government, which cannot appeal; and it proves to the political and corporate leadership that the mother-activists are not only credible but also dedicated witnesses on behalf of injustice.

However, this success in creating a political response outside of the country has not proven as successful in repairing the damage caused by the public woman discourse within it. Indeed, while the mother-activists' declaration of motherhood has served as the principal experience motivating their political organizations, they have not been immune to the accusation that public women have infiltrated their organizations and have contaminated their motherhood. For since the public woman discourse focuses so powerfully on the concept of contagion, the presence of "non-mother" activists within the organizations represents a constant liability via their potential ability to "contaminate" the mother-activists and turn them into public women. Each organization has been publicly excoriated for harboring public women who have turned good private mothers into the kinds of women who seek personal fortune and fame at the expense of their and other Mexican families.[36] And such accusations have frayed the alliances among the family organizations, which, in 2004, publicly parted ways around the accusation that some organizations were not dedicated to the mothers' *real* concerns but instead to the political and economic interests of feminists, international organizations, and other groups not characterized by the "family women" of Mexico.[37] This public accusation, which eventually has been directed at all of the family groups, first occurred during the coordination of the February 2004 protest in Ciudad Juárez organized in collaboration with Eve Ensler's V-Day Foundation and Amnesty International, among other local nonprofit organizations on both sides of the border.[38] One of the family organizations, Nuestras Hijas de Regreso a Casa, accused the event organizers, which also included leaders from Justicia para Nuestras Hijas, of having "sold out the event" to "American feminists." This accusation exposed the idea that the event organizers were working on "foreign" concerns or, as it were, concerns "unfamiliar" to the mother-activists. The third family organization had already left town altogether in order to avoid the event. So on the day of the march, press attention focused on the splintering of the coalitions, the spreading out of mothers' attendance at some events over others, and on the accusations of public womanhood that characterized the public fallout.[39]

In the aftermath of this controversy, the coalitions that constituted the femicide movement within Mexico fragmented, as its constituent groups focused on their own goals. The family organization in Ciudad Juárez, Nuestras Hijas de Regreso a Casa, for instance, focused its energies on international organizing as a way to generate domestic pressure, such as through the international human rights activism. But, as one of the founders, Irma Campos, of the Las Mujeres de Negro group, told me in Chihuahua City, "The femicide movement in Mexico is very weak right now. We're all working on our own projects. Some of them are important, very important, but the femicide movement is now more outside the country than inside it."[40] For instance, at the time of our conversation, Irma was working on the passage of the new domestic violence legislation that would transform the legal treatment of intrafamilial abuse and serve as a model for legislation in other states. She had been, as a Mujer de Negro, singled out by the governor's office in 2003 as being a public woman who was "profiting" from selling stories of the crimes to an international audience and who was adding to the "social disintegration" of Mexican society—this within a city where Irma and her husband are well-known political figures.[41]

CONCLUDING REFLECTIONS

A tension illustrated in the shifting strategies of the femicide movement in northern Mexico is that, with regard to mother-activism, it has proven easier to generate public empathy for the movement internationally than it has been to do so domestically. In other words, empathy has been more forthcoming from strangers than from those more personally familiar with the activists, the events surrounding the crimes, and the issues at play in the political economy of northern Mexico. This situation raises important questions for activists and scholars who work to connect the intimate experience of injustice with globalized campaigns in the name of justice. The efforts to create such circuitry, as seen in femicide activism, expose the challenges of generating activism around familiarity when such efforts prioritize the expectations of a global activist audience over those of a domestic one. If anything, this case shows how producing the circuitry through which intimate stories of pain and injustice can travel globally come with costs as well as with resources.

In order to understand the meaning of such dynamics for the femicide movement we must return to the initial question regarding the costs of establishing familiarity as a criterion for political action. When we examine this question within the context of femicide activism in northern Mexico, we can see that the need to establish familiarity as a pre-condition to creating a political movement that links activists with an activist public has been prominent throughout the country's history. The Coalition's struggle to confront the accusation that the victims had caused their own deaths by behaving like whores was a struggle to familiarize the victims by making them recognizable as "daughters" to the public. While whores are "familiar" within the discourse of the public woman that equates women in the sex industry with social trouble and contagion, such women are not the kind who are familiar as family women, or the kind of women that good families want to know. So by turning the victims into daughters, via the discourse of them as innocent children, the activists created a political movement around the public empathy for the victims as the kind of people that the public not only could know but also could love.

The strategic reworkings of the leading activist organizations and their practice of testimonial witnessing also reflect such dynamics, as they constantly had to assert the credibility of the activists as "good family women" within the iconic presentation of "motherhood." The attack on the Coalition and the domestic dissolution of the femicide alliances has not meant the end to activist politics against violent misogyny, but it has meant that domestic pressure on elected officials and corporate leaders is weaker than it was in the 1990s.[42] By contrast, the success of mother-activism in using testimonial witnessing as a strategy for internationalizing the movement and creating a political public that responds to the message of mothers who are fighting for their daughters has created real pressure on the federal government. While the meaning of the Inter-American Commission decision for Mexicans in relation to the government has yet to be seen, the potential precedence is itself noteworthy, as femicide is now, thanks to the activists from the Coalition to the mother-organizations, a political issue for which governments across the Americas can be held accountable. However, the targeting of mothers for being associated with public women and the ongoing association of feminists with socially dangerous women continues to plague the movement and hamper its ability to create a political community within the country. The public woman discourse, in short, has been an incredibly powerful tool for sowing public suspicion and for casting women who participate in democratic politics as the kind of women that good families would not want to know or publicly support.

The stakes in such politics could not be higher than they are now as northern Mexico, and Ciudad Juárez particularly, is in the throes of the worst violence in its modern history. While marches and protests have occurred, particularly with the 2011 national caravan organized with the poet Javier Sicilia, a forceful social justice movement has yet to develop in northern Mexico. The murdering of activists and academics, along with thousands from the working poor, has reached proportions hardly imaginable a few years ago, and the lack of sustained protest over the inept policing, the inability of the government to stop the violence, and the ongoing impunity of the criminals is equally stunning. Perhaps an early indicator of this terrible situation occurred with the discovery in 2003 of twelve men's bodies, all showing signs of torture, buried in the backyard of a middle-class home, just a few kilometers from where the eight female bodies had been found in 2001. This discovery was not accompanied by any public protest or marches. Nor was there a public outcry over the impunity. Instead, the victims were all declared to be *narcos* (drug gang members), the house a *narco-casa*, the mass grave a *narco-fosa*, and the bodies evidence that the victims had invited the trouble that had ended their lives. Today, there is still much to be done in order to familiarize the victims of drug violence to a public that sympathizes with them and that acts on their behalf politically. No one, in other words, has succeeded in turning *narcos* into innocent *hijos*.[43] And it appears that Esther Chávez spoke with tragic prescience when she declared early in her activist days that the violence and impunity that had created the conditions of femicide would one day "turn this city into a horror."

As recent theorists, many of them feminists, have urged, such events call for a serious recalculation of the need for familiarity as a means to generate political empathy. The question, as posed by Judith Butler, is one that is central to human rights activists and theorists as they ask the potential public for political causes to suspend their need for identification with the victims and/or their witnessing activists in order to respond politically to their needs for political pressure, money, or other kinds of support. As the above events illustrate, such a recalculation is urgent for the cultivation of broad-based alliances that have the resources for expanding their political bases across geographic scales. And it is an urgency that requires willingness on the part of witnessing publics to move beyond the politics of recognition and not demand "familiarity" as a precondition for political action. Witnessing publics, in other words, must recognize their proactive roles in establishing the terms through which politics take shape and justice is demanded through the intimate encounters of global activism.

Such a move requires a commitment to what Susan Buck-Morss calls a "radically open communication" for expanding political possibilities through an expansion of the way we tell and listen to politically testimonies. Such an expansion is vital for creating a politics in which public women stand aside private ones as they fight together against violence and impunity and as they try to make their governments more accountable to their citizens.[44] Making the expansion and resilience of such alliances the explicit goal of scholarly reflection is a necessary step for fortifying the bonds of political activists and their political publics. And if the lessons of Mexico's femicide activism are to be taken to heart, a progressive politics of global justice must reflect critically on the circulation of intimacy and the costs of familiarity within the work and networks of social activism.

NOTES

In memory of Esther Chávez Cano, courageous activist, shameless feminist, dear friend.

1. Sara Ahmed, "Affective Economies"; Gillian Whitlock, *Soft Weapons*; Geraldine Pratt, "Circulating Sadness."
2. Julia Monárrez Fragoso, "Feminicidio sexual serial en Ciudad Juáres"; Amnesty International, *Muertes intolerables*.
3. A 2010 report by the Washington Office on Latin America (WOLA) presents data that fewer than 25 percent of crimes in Mexico are reported and that fewer than 2 percent result in sentences. See Meyer, "Abused and Afraid in Ciudad Juárez."
4. Anne Cubilié, *Women Witnessing Terror*; Kelly Oliver, *Beyond Witnessing*; Melissa W. Wright, "Justice and the Geographies of Moral Protest."
5. Nora Strejilevich, "Testimony."
6. See Kay Schaffer and Sidonie Smith, *Human Rights and Narrated Lives*; Cubilié, *Women Witnessing Terror*; Oliver, *Beyond Witnessing*.
7. Joan B. Landes, ed., *Feminism*.
8. See also Tamara Neuman, "Maternal 'Anti-Politics' in the Formation of Hebron's Jewish Enclave"; Geraldine Pratt, "Circulating Sadness"; Cynthia Bejarano, "Las Super Madres de Latino America"; Diana Taylor, *Disappearing Acts*; Sonia Alvarez et al., *Cultures of Politics/Politics of Cultures*.
9. Esther Chávez Cano, interview with author, Ciudad Juárez, February 2007.
10. Ibid.
11. Marta Estela Pérez García, "La coordinadora en pro de los derechos de la mujer."
12. Esther Chávez Cano, interview with author, Ciudad Juárez, November 2006.
13. Esther Chávez Cano, interview with author, Ciudad Juárez, May 2003.
14. Debra A. Castillo et al., "Border Lives."
15. I would like to thank Dr. Socorro Tabuenca Córdoba for pointing out this linguistic difference.
16. Linda Diebel, "Macabre Murders Bewilder Mexicans"; see also the online article by Gabriela Minjares and Sandra Rodríguez, "The Facts behind the Myth."
17. Anonymous, interview with author, Ciudad Juárez, May 2003.
18. Wright, "Justice and the Geographies of Moral Protest."
19. Ibid., 8.

20. I have discussed this issue at length elsewhere: Melissa W. Wright, "Urban Geography Plenary Lecture."

21. Anonymous, interview with author, Ciudad Juárez, 2004.

22. See, for instance, Sam Dillon, "Rape and Murder Stalk Women in Northern Mexico."

23. Robert Urrea, in an interview by John Quiñones, "Silent Screams."

24. For a synthesis of some of these accusations from the Ciudad Juárez dailies *El Diario de Ciudad Juárez* and *El Norte de Ciudad Juárez*, see the following article on the *Frontera-NorteSur* Web site: "Possible Trouble at Crisis Center Shows Trouble in Ciudad Juárez Press."

25. See Debbie Nathan, "The Missing Elements."

26. Kathleen Logan, "Personal Testimony."

27. Nora Strejilevich, "Testimony"; George Yúdice, "Testimonio and Postmodernism."

28. As one Argentinian activist, author, and torture survivor has written, testimony communicates the "intimate, subjective, deep dimension of horror" (Nora Strejilevich, "Testimony," 701) that the testifier has experienced personally and that reflects the experience of many.

29. See Stephen Schwartz, "A Noble Prize for Lying."

30. Sonia Alvarez et al., *Cultures of Politics/Politics of Cultures*.

31. See Kay Schaffer and Sidonie Smith, *Human Rights*.

32. See Nora Strejilevich, "Testimony."

33. Anonymous, interview with author, Ciudad Juárez, January 2007.

34. Frontera NorteSur, "International Court Holds Mexico Accountable for Femicides."

35. Ibid.

36. L. Sosa, "Recriminan a ONGs madres de asesinadas"; R. Chaparro "Se reúnen autoridades con madres de víctimas"; C. Guerrero and G. Minjares, "Hacen mito y lucro de los femincidios."

37. Melissa W. Wright, *Disposable Women and Other Myths of Global Capitalism*.

38. Clara Rojas, "The V Day March in Mexico."

39. Melissa W. Wright, "Public Women, Profit, and Femicide in Northern Mexico"; David Piñon Balderrama, "Lucran ONGs con muertas."

40. Irma Campos, interview with the author, Chihuahua City, February 2007.

41. Benjamin Martínez Coronado, "Desintegración sociofamiliar, Germen de Crímenes."

42. Wright, "Justice and the Geographies of Moral Protest."

43. At the time of this writing, some activists are taking on this challenge in the aftermath of a massacre of children at a birthday party on January 31 and in response to the President's declaration that the violence reflected violence "between rival gangs." See Gabriela Minjáres, "Reprimen manifestación de estudiantes."

44. Susan Buck-Morss, *Thinking Past Terror*.

BIBLIOGRAPHY

Ahmed, Sara. "Affective Economies." *Social Text* 79 (2004): 117–39.

Alvarez, Sonia, Evelyn Dagnino, and Arturo Escobar. *Cultures of Politics/Politics of Cultures: Re-visioning Latin American Social Movements.* Boulder: Westview Press, 1998.

Amnesty Internacional. *Muertes intolerables—México: 10 años de desapariciones y asesinatos de mujeres en Ciudad Juárez y Chihuahua.* London: Amnesty International, 2003.

Balderrama, David Piñon. "Lucran ONGs con muertas." *El Heraldo de Chihuahua,* 23 February 2003.

Bejarano, Cynthia. "Las Super Madres de Latino America: Transforming Motherhood by Challenging Violence in Mexico, Argentina, and El Salvador." *Frontiers: A Journal of Women Studies* 23, no. 1 (2002): 126–50.

Buck-Morss, Susan. *Thinking Past Terror: Islamism and Critical Theory on the Left.* New York: Norton, 2003.

Castillo, Debra A., Maria Gudelia Rangel Gomez, and Bonnie Delgado. "Border Lives: Prostitute Women in Tijuana." *Signs* 24, no.2 (1999): 387–433.

Chaparro, R. "Se reúnen autoridades con madres de víctimas: Lucran ONGs con feminicidios, acusan." *El Diario de Ciudad Juárez,* 22 January 2005.

Coronado, Benjamin Martínez. "Desintegración sociofamiliar, Germen de Crímenes: Patricio." *El Heraldo de Chihuahua,* 20 February 2003.

Cubilié, Anne. *Women Witnessing Terror: Testimony and the Cultural Politics of Human Rights.* New York: Fordham University Press, 2005.

Diebel, Linda. "Macabre Murders Bewilder Mexicans: More than 100 Women Slain near Border Since 1993." *Toronto Star,* 7 December 1997, A1

Dillon, Sam. "Rape and Murder Stalk Women in Northern Mexico." *New York Times,* 18 April 1998.

Fragoso, Julia Monárrez. "Feminicido sexual serial en Ciudad Juáres: 1993–2001." *Debate Feminista* 25 (2002): 279–303.

FronteraNorteSur. "International Court Holds Mexico Accountable for Femicides." *New America Media,* 15 December 2009. http://news.newamericamedia.org/news/view_article .html?article_id=f6cf9510252873325ffe81c487a02fdf.

——. "Possible Trouble at Crisis Center Shows Trouble in Ciudad Juárez Press." *FronteraNorteSur,* February 2001, www.nmsu.edu/~frontera/feb01/hmrt.html.

Guerrero, C., and G. Minjares. "Hacen mito y lucro de los femincidios." *El Diario de Ciudad Juárez,* 22 July 2004.

Landes, Joan B., ed. *Feminism: The Public and the Private.* New York: Oxford University Press, 1998.

Logan, Kathleen. "Personal Testimony: Latin-American Women Telling Their Lives." *Latin American Research Review* 32, no. 1 (1997): 199–211.

Meyer, Maureen, with contributions from Stephanie Brewer and Carlos Cepedai. "Abused and Afraid in Ciudad Juárez: An Analysis of Human Rights Violations by the Military in Mexico." Washington Office on Latin America (WOLA), September 2010. Available at

http://idpc.net/sites/default/files/library/Abused%20and%20afraid%20in%20ciudad%20
juarez.pdf

Minjáres, Gabriela. "Represent manifestación de estudiantes." El Diario de Ciudad Juárez,
12 February 2010.

——, and Sandra Rodríguez. "The Facts Behind the Myth." *El Diario de El Paso,* www.diari-
ousa.com/news/specials/juárez_women_murders/ (accessed 8 February 2010).

Nathan, Debbie. "The Missing Elements." *Texas Observer,* 30 August 2002.

Neuman, Tamara. "Maternal 'Anti-Politics' in the Formation of Hebron's Jewish Enclave."
Journal of Palestine Studies 33, no. 2 (2004): 51–70.

Oliver, Kelly. *Beyond Witnessing.* Minneapolis: University of Minnesota Press, 2001.

Pérez García, Marta Estela. "La coordinadora en pro de los derechos de la mujer: Política y pro-
cesos de cambio en el municipio de Juárez (1994–1998)." Master's thesis, La Universidad
Autónoma de Ciudad Juárez, 1999.

Pratt, Geraldine. "Circulating Sadness: Witnessing Filipina Mothers' Stories of Family Separation." *Gender, Place and Culture* 16, no. 1 (2009): 3–22.

Rojas, Clara. "The V Day March in Mexico: Appropriate and (Mis)Use of Local Women's
Activism." *National Women's Studies Association Journal* 17 (2005): 217–28.

Schaffer, Kay, and Sidonie Smith. *Human Rights and Narrated Lives.* New York: Palgrave
MacMillan, 2004.

Schwartz, Stephen. "A Noble Prize for Lying." *Globe and Mail,* 30 December 1998.

Sosa, L. "Recriminan a ONGs madres de asesinadas." *El Diario de Ciudad Juárez,* 6 July 2007.

Strejilevich, Nora. "Testimony: Beyond the Language of Truth." *Human Rights Quarterly* 23,
no. 3 (2006): 701–13.

Taylor, Diana. *Disappearing Acts: Spectacles of Gender and Nationalism in Argentina's "Dirty
War."* Durham: Duke University Press, 1997.

Urrea, Robert, in an interview by John Quiñones. "Silent Screams: Who Is Killing the Young
Women of Juárez?" *20/20,* ABC, 20 January 1999.

Whitlock, Gillian. *Soft Weapons: Autobiography in Transit.* Chicago and London: University
of Chicago Press, 2007.

Wright, Melissa W. *Disposable Women and Other Myths of Global Capitalism.* New York:
Routledge, 2006.

——. "Justice and the Geographies of Moral Protest: Reflections from Mexico." *Environment
and Planning D: Society and Space* 27 (2009): 216–33.

——. "Public Women, Profit, and Femicide in Northern Mexico." *South Atlantic Quarterly* 105,
no.4 (2006): 681–98.

——. "Urban Geography Plenary Lecture: Femicide, Mother-Activism, and the Geography of
Protest in Northern Mexico." *Urban Geography* 28, no. 5 (2007): 401–25.

Yúdice, George. "Testimonio and Postmodernism." *Latin American Perspectives* 18, no. 3 (1991):
15–31.

14

SOLIDARITY, SELF-CRITIQUE, AND SURVIVAL
SANGTIN'S STRUGGLES WITH FIELDWORK

Sangtin Writers
Reena, Richa Nagar, Richa Singh, Surbala

PROLOGUE

In 2002, nine women began a collective journey in the Sitapur District of India's Uttar Pradesh as activists, critics, writers, and close companions. Eight of these women were grassroots workers in the district office of Mahila Samakhya Programme (MS), a large and influential government-sponsored nongovernment organization (NGO). The ninth member was an academic in a U.S. university who also worked as a creative writer and volunteer in people's organizations in Uttar Pradesh. The journey sought to challenge the traditional separations among activist, academic, and creative labor by evolving methodologies through which grassroots leadership could emerge to collectively rearticulate the needs of the rural poor and to create the conditions under which rural poor could organize to meet those needs. By building solidarities across the divides of scholarship, creative work, and grassroots organizing, the collective scrutinized—through dialogue, autobiographical writing, and critical analysis—the successes and limitations associated with projects that seek to empower poor rural women. Our goal was to challenge hierarchies that marginalized the perspectives of grassroots workers and to then develop processes that could empower them to articulate their own vision for sociopolitical change. In this way, Sangtin, an organization founded by the village-level membership of MS-Sitapur in 1999, hoped to continue the work of self-empowerment in Sitapur's villages after the time-bound funded program of MS had "rolled back" from that district. (For the meaning of *sangtin* and other Hindi and Awadhi words, see the glossary following the article.)

This journey led to the collectively authored book in Hindi, *Sangtin Yatra* (2004). The book's critique of donor-funded empowerment projects ignited a series of criticisms and responses that we then chronicled in the book's English version, *Playing with Fire* (2006).[1] It was from these intellectual battles that Sangtin Kisaan Mazdoor Sangathan (SKMS) evolved in 2005. SKMS resolved to grow as a people's movement without depending on donor funds. Donor groups typically target women in isolation from the men in their communities. SKMS rejected this approach as well as the assumption that violence against women could be addressed separately from the violence generated against rural livelihoods by everyday sociopolitical and economic processes.

The renaming of Sangtin as Sangtin Kisaan Mazdoor Sangathan and the emergence of SKMS as a principally member-supported movement[2] went hand in hand with the mobilization of 2,000 women and men from sixty villages to restore the flow of irrigation waters to a distributary channel that had been written off as unusable for sixteen years. During this 2005 campaign, SKMS also began to address the issue of nonpayment of minimum wages and the non-inclusion of *Dalit* ("untouchable") women in specific forms of formal labor. Sitapur was one of the 200 pilot districts in which India's National Rural Employment Guarantee Scheme (NREGS) was launched in 2005. SKMS pushed for its fair implementation, building a sustained agitation to ensure fair access to work and adequate wages. As the movement grew, SKMS stepped into the politically volatile terrain of demanding unemployment benefits for those who had not found work. This campaign for unemployment benefits, launched in November 2007, resulted in a major victory in February 2009 with 826 households in Sitapur winning unemployment benefits under NREGS.

This chapter is based on excerpts from SKMS's forthcoming book in Hindustani, *Ek Aur Neemsaar*.[3] *Ek Aur Neemsaar* embraces some of the key intellectual and political challenges articulated in *Sangtin Yatra/Playing with Fire*, and analyses the processes by which the public debates generated by *Sangtin Yatra/ Playing with Fire* fed into the making of a movement of 5,000 peasants and laborers, and how this movement has tried to provide an ongoing critique of mainstream projects of empowerment while also pushing the state to implement its own policies adopted in the name of "the poorest of the poor." We examine the complex politics of alliance work at local, national, and transnational scales, and the ways in which the pasts and presents of systematic humiliation, deprivation, and disability shape the making of a movement. In so doing, we seek to rework messy questions surrounding the global politics of knowledge production—including "fieldwork" as practiced by both academics and NGO-actors—through a concrete engagement with ghettoization of women's issues; a collective theorizing of caste-, class-, and gender-based violence in rural India that has been thoroughly subjected to neoliberal policies; and the contradictions and possibilities embedded in intellectual and political alliances across difficult sociopolitical borders.

1

Once, a scholar who considered himself highly learned and accomplished climbed on a boat. The simple boatman respectfully welcomed the *pundit*. As the boat start sailing, the *pundit* asked the boatman, "Hey, do you know anything about capitalism?"

The boatman folded his hands, "I am a sheer illiterate, sir. How would I know about capitalism?"

"What's the point of living in such darkness? You have wasted one-fourth of your life," the learned man pronounced as he tossed some *pan-masala* in his mouth.

14.1 SKMS's hunger strike in Sitapur's Vikas Bhavan (district development headquarters), May 2008

As the boat sailed further, the *pundit* was once again taken by an urge to establish his intellectual authority—"So, you are a laborer. I am sure you have heard of Marxism!"

"Where would I learn about that, *sahab*? I know nothing of it."

"What a pity. You have ruined half of your life! But you are married, aren't you? Don't tell me you haven't heard of feminism!" the scholar teased.

When the boatman expressed his ignorance of feminism, too, the *pundit* declared three-quarters of the boatman's life wasted. Before the boatman could respond to this declaration, the boat began to sink in midstream. The boatman said to the scholar, "*Pundit-ji*, you know everything. Now, swim."

But alas, the *pundit* had not learned to swim.

"*Pundit-ji*, I have merely wasted three-quarters of my life, but you are just about to lose all of yours!" Saying this, the boatman quickly swam across the river. The scholar drowned.

This story can often be heard from the mouths of the members of SKMS in the villages of Sitapur. And the story is often followed by the question: "Why didn't the boatman try to save the scholar?" If the scholar had not incessantly mocked the boatman and belittled his knowledge and existence, there is a good chance that the boatman would have extended his hand to save the scholar, and perhaps the knowledges and lives of both the men could have been saved.

If we agreed that each one of us is part *pundit* and part boatperson, there would be very little to hold in place the walls that separate intellectuals from the authors of everyday lives and struggles. Unfortunately, however, the world that we live in is inflicted with a system in which most of us do not get the opportunities to develop both of these elements within us, and the gaps between the *pundit* and the boatperson become a vehicle to reinforce the differences of caste, class, gender, race, and place.

Who becomes the subject of knowledge and who is designated as the *pundit* to produce, legitimate, and disseminate that knowledge? And what are the implications of this inequality? Raising these questions in the context of the politics of rural development and women's empowerment, nine *sangtins* began a journey that first acquired the form of the book, *Sangtin Yatra*, and subsequently evolved into a movement that now comprises 5,000 workers and peasants, both women and men, more than 90 percent of whom are *Dalit*. Grappling with the nuances of class, caste, gender, and communal differences in relation to "Development," this journey is, on the one hand, committed to securing livelihoods and the right to information and to bringing waters to a dry irrigation canal for the members of the movement. On the other hand, in linking the issue of socioeconomic disempowerment of the poor with their intellectual disempowerment, this journey poses the question of who determines target populations, issues, and activities for projects of empowerment, and how? Who keeps the accounts of these activities and for whom? And how do the coordinators of these projects, in becoming the "experts" on the disempowered, contribute to maintaining the status quo?

As SKMS members go through their own soul-searching on these issues, we also experiment with ways in which connections between intellectual labor and grassroots organizing can be deepened, and how the realities of political, social, and intellectual hierarchies, as well as the very definitions of who and what constitutes the margins, can be complicated and transformed.

It was as a part of these experiments that the idea of inviting a few *sangtins* to the United States emerged in 2005. Richa Nagar first shared this idea with Rajani (pseudonym), a volunteer in a U.S.-based organization that considers itself a friend of SKMS. Rajani welcomed the proposal, and as she warned us about the visa-related challenges, she also posed a difficult question—"If only two or three women came here, how would this impact the rest of the organization? How would you explain to your members that some of them are good enough for a trip to the U.S. and others are not?"

Rajani's curiosity was linked to the challenges of creating equitable distribution of opportunities and platforms in an emerging organization; members of SKMS were also having serious discussions on this subject. Rajani's way of raising this important question, however, betrayed a lack of critical self-reflection. Members of SKMS asked: When so many well-intentioned professionals and social workers from organizations in the United States land in places such as Sitapur without an invitation, does anyone ask them how their presence would affect whom and where? Are members of SKMS incapable of assessing for themselves the effects that the traveling of a few representatives of our organization to the United States could have on the dynamics within the movement?

But how often are organizations from the south able to pose these frank questions before philanthropic or academic visitors from the global north? Perhaps much of the third world often begins to appear to these well-wishers as their own "field." These visitors often presume that they can plant their own tent wherever they arrive, and that the dwellers of surrounding villages and slums would welcome them and their expertise and charity with open arms. Those of us who are in Rajani's situation travel easily from one world to another and such travel is readily accepted as part of our work, or as our attempt to deepen our learning or to gain necessary exposure. However, if a village-level activist seeks a similar opportunity, experts such as Rajani can subject the idea to scrutiny without even inviting organizational workers from the global south to be part of the dialogue.

Along with the goal of strengthening SKMS in Sitapur, it also seemed important to challenge the notion that it is only first world experts who must travel in order to carry out development fieldwork. Third world people in struggle, who are desirous of sharpening their understandings of the context of development and the related politics of funds, knowledge, and languages, must also make the first world their own field. In so doing, they can generate dialogues on all those complex issues, the rights to deliberate about which are often claimed by the world's esteemed degree-holders on behalf of the degreeless poor and rural majority. Turning this desire into reality was facilitated by *Playing with Fire*. Thus materialized another segment of Sangtin's journey—in a series of events in Syracuse, Minneapolis, Stanford, Berkeley, and San Francisco. The churnings that accompanied and followed this fieldwork were marked by tensions between solidarity and self-critique, especially in the context of the changing configuration of the movement, and attempts to grapple with the ways that processes associated with NGOization continue to haunt our own relationships even as the movement works to confront these very processes. Key issues include questions of monetary compensation and the contradictions arising between the labor of movement building and the need for stable livelihoods. We turn to some of these issues in the next section.

2

The book release of *Sangtin Yatra* in Lucknow in March 2004 marked a new phase in our journey. Even before the book sparked controversy for its critique of NGO-driven empowerment, it gave us a glimpse of the difficult turns that were awaiting us. Having learned to recognize and confront the politics of caste, class, and communalism in the context of NGOs, we were now being pushed into another arena—of recognition, limelight, and egos—that was marked by new political puzzles: Who wants to hold whose hand at which juncture in a struggle? When and how do certain "hand holders" want to be recognized for holding the hand? Several NGO officials, writers, and academic critics repeatedly posed a familiar string of questions before us: Which individuals should be credited or blamed for the success of *Sangtin Yatra*? Whose words were they, after all? Whose analysis was it? And whose courage? Which of the nine authors should be crowned for its success and which ones should be placed in the stand and indicted? Was the collective spirit of the book a lie?

At first, the recurrence of these questions tormented us. But we soon learned to embrace these questions as a necessary part of refining our own critique of NGOization—our understandings of how ideas, successes, and celebrity were circulated and recognized in a local, national, and global market, and how we had to continuously define our relationships to these processes without pretending that we could sidestep them. On the surface, the difficult dialogues that ensued from this process reflect the story of the building of a movement. But if one digs deeper, one can see many stories—of the building, crumbling, weakening, and stabilizing of relationships; of welding together different worlds to continue a struggle; of an alliance's efforts to find continuous hope in circumstances that often generate horrifying hopelessness.

Even as this dialogue—of relationships, struggles, and hopes—makes us attentive to never-ending battles for livelihoods and dignified existence, it enables us to identify more sharply the contours of our battlefield by grappling with difficult questions of representation, mediation, and exploitation, and with the eternal need for unrelenting self-critique. Thus, when we talk about the rapid growth in SKMS's membership, we have no hesitation in acknowledging that the toughest labor of deepening the critical analysis of the organization's work and its directions has only begun. And as we smile at the memory of SKMS's Holi meeting of 2007, when two *Sawarn* men touched the feet[4] of all the *Dalit* women for the first time in the history of Sitapur, we also see flashing before our eyes the faces of those *Sawarn* members who refuse to eat or drink in a *Dalit* member's home or who cringe at the thought of their daughters running away and marrying Muslim men.

While it is easy to discuss and celebrate collectivity in a rapidly growing movement, we cannot shy away from the task of critically assessing how the form of our collectivity is changing, and the manner in which these changes produce tensions and conflicts. Thus, as several thousand new co-travelers join us we must also understand why some old ones had to abandon the journey after we had traveled many miles together. For it is precisely a responsible analysis of this parting and merging of ways that allows SKMS to grapple with tension-ridden, yet tightly intertwined relationships among solidarity, movement-building, and NGOization.

The writing of *Sangtin Yatra* was like creating our constitution. The critique of NGOization spelled out in that constitution led the director of MS-Uttar Pradesh to take disciplinary action against the seven authors who worked in

14.2 Reena addressing a rally

MS-Sitapur. Of these seven, Richa Singh resigned from MS rather than accept a transfer, but the collective, with support from allies, successfully fought to save the jobs of the other six—Anupamlata, Reshma Ansari, Ramsheela, Shashibala, Shashi Vaish, and Vibha Bajpayee.[5] As Richa Singh, Surbala (who had quit MS in 2001), and Richa Nagar committed themselves to transforming into reality the dream we had begun to dream in *Sangtin Yatra*, the remaining six writers took the hard decision of maintaining an intellectual connection with the struggle but without immersing themselves in the everyday labor of building an organization. After all, a dream woven through a book was not enough to provide *daal-roti* to a family! Four of these authors subsequently received promotions in MS and accepted positions outside Sitapur. In some ways the promotions can be seen as an acknowledgment by MS of the importance of their intellectual labor in (and of their visibility following) *Sangtin Yatra*.

The ideas woven together in *Sangtin Yatra* had emerged from dialogues among nine women. But those ideas remained a mere skeleton in the absence of real flesh. That flesh could materialize only by continuous walking, building, learning, and doing. As the six authors employed by MS became immersed in their own new goals and lives, they found it hard to participate in such a process; their new responsibilities made it increasingly difficult for them to take out time for the next phase of the *yatra* ("journey"). Fortunately, three other employees and volunteers associated with MS—Reena, Geeta, and Kshama—quit MS and became full-time volunteers in organizing this next phase of the journey.

Those of us who were absorbed in building a new organization found it tough to remain deeply connected with our old companions in the face of the intense demands of building a movement. In the middle of the first campaign around access to irrigation, Geeta and Kshama also chose to part ways with us when they did not see any hope for salaried employment; they took regular jobs in local NGOs. Their departure was profoundly saddening, but the words of one of our mentors helped us to deal with this loss: "The *yatra* of *sangtins* is no different than a long, long journey by train. At every major station, you should expect some travelers to climb aboard and others to descend."

This metaphor of a long train ride worked like a *mantra* for us. It gave us a language to detach ourselves from the idea of "the original authorship" of *Sangtin Yatra* and to build new relationships that would fire our movement. As more and more peasants and laborers began to submerge themselves deeper into the colors of *Sangtin Yatra*, the colors spread, brightened, merged, and found new hues and shades. Those who grew distant from the process remained untouched by these new colors. They retained only our old color, one that was bound to fade and stand apart as it blended with the new colors brought by the new shades in their own journeys.

Some authors of *Sangtin Yatra* argued that their status as founding members of the movement made them different from the new membership of SKMS on the grounds of education and qualification; therefore, SKMS should seek funds to create stable employment for all the writers of *Sangtin Yatra*, and their salaries should, at the very least, match what they were making in MS. However, the 5,000 members who have built SKMS since 2005 have made a policy decision that we must continuously work to build an antihierarchical organization that makes no distinction between the founders and new members. This policy includes an unwritten agreement that anyone desirous of being a part of the movement should be prepared to devote significant energies as an unpaid volunteer for at least six months without any expectation of monetary compensation. One of the most difficult and decisive meetings of the authors' collective took place in July 2007 after SKMS decided to give minimum wages by rotation to each of its volunteers who had to leave their farm work or daily wages to do the organization's work. This decision was not acceptable to several authors of *Sangtin*

Yatra. Two asked, "When we built the foundation of a house, how is it that those who arrived after us have found shelter in the house, but no one is guarantying us a roof?"

Such conflicts forced us to distinguish between the authors of the book, *Sangtin Yatra*, and the authors of everyday struggles who have labored and hungered to make SKMS a powerful movement. And we return once again to the metaphor of the train stations. We all had planned to board the train and to travel from one village to an unknown place, but a few members chose to leave the train in the middle of the journey, while the others managed to continue the journey, on which they were joined by several thousand new companions. It is true that those who left in the middle were critical in undertaking the journey, but if they want to join the action at the current location, there is no shortcut for them. They will have to find a way to traverse the distance they have missed.

3

We were in Syracuse in Chandra Talpade Mohanty's graduate seminar, and before the discussion on *Playing with Fire* began, a woman called Andeline came and hugged us. Tears shimmered in her eyes as she remarked: "Your book sent me back to the children with whom I lived before coming to the U.S." Later we learned that Andeline had worked with Namibian children living with HIV/AIDS. In the childhoods of *sangtins* she saw the childhoods of those children, and the connections came to her not in the form of words, arguments, or definitions, but as tears. So much power was there in Andeline's tears that as the discussion about the book grew deeper, several students in the class began to cry. None of us had ever before encountered such a fearless embracing of tears in an academic setting. This raised a powerful question in the discussion: Which forms of knowledge and expertise remain unrepresented when tears do not find a space in intellectual conversations? Instead of assuming that knowledge can prevail only by refuting, contradicting, and refining an argument with an even sharper one, can we not present our arguments effectively through the language of tears, silences, and relationships?

When Richa Singh and Surbala traveled to Minnesota, Richa Nagar wanted them to meet those friends who had taken an interest in our journey and offered their support at moments of crisis. About thirty-five such friends and supporters came over for a potluck lunch, and the afternoon was filled with conversations, connections, laughter, and delicious food. Hours later, at two in the morning, we talked as we looked over the lights along the Mississippi River from Richa's seventeenth-floor apartment. Watching the scene, Richa Singh's eyes grew wet. So much food during the day and now these glittering lights reminded her of Tama's old mother, who stands in a long queue outside the ration shop with a tiny bottle so that she can borrow enough oil to cook the next day's meal, and many times she returns home empty-handed. It is not as if glittering lights do not exist in Delhi or that there are no discussions about hunger and poverty in the midst of parties in India. But being together in the United States allowed us the time and space to confront these contradictions in the context of our journey and alliance.

On one level, we collectively recognized the gaps that separate Tama from the platforms where people analyze the complexities of Tama's *Dalit* identity, his blindness, or his family's hunger. On another level, we were also able to understand that gulfs exist among us in innumerable ways—sometimes they take the form of Richa Nagar's status as a U.S.-based professor and a person from a well-known literary household in India, which separates her from all the other members of the movement; sometimes they can be seen in the distance that separates Richa Singh's world in Sitapur town from the world of Satnapur, where Surbala returns home every night; sometimes they acquire the form of the socioeconomic rifts that stand between the families of Surbala and Tama; and sometimes they take the form of Tama's own upbringing and stubborn values that prevent him from eating in the home of a *Raidas*.[6] The only way to advance through these contradictions is by tirelessly confronting them.

In Minneapolis, members of Ananya Dance Theater invited Surbala, Richa Singh, and Richa Nagar for a discussion. The dancers, several of whom are feminist academics, were keen to discuss the same point in *Playing with Fire* that several other readers of color in the United States had wanted to delve into—the moment when Radha tells Sandhya that there is a huge difference between feeling someone else's pain and living that pain. When asked to elaborate on this difference between living and feeling, we argued that Radha is saying that the struggle of living an injury is far greater than feeling another's pain. But she is not saying that those who can only feel another's pain of exploitation or humiliation cannot stand in solidarity with those who live that pain. Nor is she claiming that those who live an injury are always in a position to identify it as such. In other words, living the wounds of injustice and the ability to feel the pain of that injury are both necessary pieces of a movement's vision and work. Those who suffer an injury have to recognize it as an injury in order to fight against it, and those who wish to stand in solidarity must develop the ability to recognize and feel the pain of that injury in order to truly become part of the struggle.

To clarify the point, Richa Singh narrated a story she had heard from her father. In a village of *Sawarns*, Qadir Mian, a Muslim man, used to frequently visit the home of his close Hindu friend, Phoolchand, where Qadir's eating and drinking utensils were always kept far away from those of the others. One day Qadir Mian was sitting in Phoolchand's backyard, and a dog was happily licking a *kadaha* right before his eyes. It is quite common for a dog to lick a *kadaha* in the villages of Uttar Pradesh, and it is just as common for sugarcane juice to be cooked in that *kadaha* after it has been licked by a dog. But Qadir Mian, who had silently lived the insult of communal untouchability for years, suddenly felt the pain of that insult. Rising from the *Khatolia*, he said, "From today, I refuse to step into a home where people accept *gur* that is cooked in a utensil licked by a dog, but where a human being following another religion is deemed untouchable." After this, Qadir Mian did not cross the boundary of Phoolchand's home until his dying day.

Translating struggles from one world to another is not an easy task. Such translation requires creating languages—through new images, metaphors, and symbols. Simple stories have an amazing capacity to narrate extremely complex struggles across worlds. Our discussions in the United States often inspired us to remember the story of a bird who tries to save her chicks from a massive fire by drawing drops of water into her tiny beak and pouring them one after another over the fire. The bird knows that the brutal fire of globalization will eventually turn her nest into ashes, but she continues to bring water in her beak in a simple hope: that when an honest history of the world is written, she can be counted among those who tried to extinguish the fire and not among those who ignited it.

4
_____

> *We were headed to Syracuse. As soon as I walked through the security check, the machine shrieked in the same way that it had done in New Delhi and Amsterdam. The culprits were my bangles and anklets. . . . The anklets came off, but the glass bangles with iron rims were impossible to take off—they had been on my wrists for nineteen years. I passed the check point after the usual drama, but for the rest of the trip I kept worrying, "How often will I have to deal with this? These bangles will draw me in the circle of suspicion each time we travel somewhere."*
>
> *The bangles became a barrier in my way—just like those stubborn questions that become thorns in my heart until I resolve them. . . . I recalled in*

that instant how once a gender-trainer from the city interrogated us in MS for putting sindur in our hair. Sindur, she claimed, was symbolic of patriarchy. In some branches of MS, village-level workers were prevented from putting on sindur. But in Sitapur, we refused to comply with this expectation because we believed that it was a form of ideological exploitation. Sindur and bangles give us tremendous respect in the places we come from. So why should we give these up simply because they are deemed unfeminist by our gender-trainers?

—FROM SURBALA'S U.S. DIARY, APRIL 2007

In Syracuse, Surbala cut her iron-rimmed bangles after nineteen years and left the pieces in the hotel's trash. Someone who was a witness to all this could argue, perhaps, that it was easy for Surbala to get rid of her bangles because she was in a place where her bangles carried no meaning. But for Surbala, whose identity as a woman was intertwined with these bangles for almost two decades of her marriage, it was a huge mental struggle to convince herself that the removal of bangles was critical for her to make the rest of her path smoother. If someone had pressured her to remove her bangles and anklets because these would cause problems along the way, Surbala would have never removed them. Even if she had, she would have considered such a pressure as an "ideological exploitation." As Surbala points out, "Arriving at the conviction that those bangles had to be cut and discarded had to be my own journey. In pretty much the same way as a butterfly must find its own strength to break its cocoon. Only then, will the butterfly be strong enough to reach its full flight."

Unwanted assistance in breaking its cocoon can be a curse for the butterfly.

Those who swear by the label "feminism" as well as those who trash it are often guilty of equating feminism with an already cooked recipe of ideas and positions. Many of these people assume that feminist practices can be widely found among educated, urban, salaried women and that these women are capable of empowering another category of women who are *Dalit*, exploited, and subordinated. What are the issues of these supposedly subordinated women? How must they be empowered? How should they form their relationships with their lovers, husbands, mothers-in-law, sons, and daughters? The solutions associated with these questions often sit in canned forms on all the so-called feminist shelves. Expert feminists will not be able to listen to our movements unless they are prepared to throw out these canned assumptions en masse and restock their shelves with a new approach to feminism. Our companion Kamala, from the Khanpur, village finds it imperative to resist the language of those women's NGOs, which repeatedly tell her to prove her credentials as an "empowered" woman: "If you are truly empowered, go fight your man for your half of the bread!" Kamala retorts: "Which half of my *roti* should I fight for when my family's access to any *roti* at all is gravely endangered? Fighting for half a *roti* is meaningless for me in the absence of a struggle for the whole *roti*."

In the United States, some of our most productive conversations happened with activists working on immigrant labor rights. Many of them were themselves immigrants from Central America and East Africa, including Somali hotel workers and Mexican farm workers, who have made the United States their temporary or permanent homes, and who are fighting for better wages, health care, and education for their companions. Their children want the same opportunities for education that are available to all U.S. citizens on paper, and they frequently draw their inspiration from the many battles that are being waged by the people of the Americas. One such battle is that of the teachers of Oaxaca in Mexico, about which we saw the film *Granito de Arena* ("Grain of Sand"). In this film, which documents a courageous movement of rank-and-file teachers opposing the privatization of education, the Uruguayan writer Eduardo Galeano remarks at the end: "There is an old proverb that says it's better to teach someone to fish than to give them fish. But what if they sell the river or what if they poison the river? And what good is it to know how to fish if the owner of the river doesn't let us fish?"

Whether it is an NGO or the academy, everywhere people seem to be awaiting or planning the next "project." When we write a book, the professionals ask us, "What will your next book be?" When we start fighting for the fair implementation of the National Rural Employment Guaranty Scheme, we are asked by NGOs, "What's your next project?" But the battles for livelihoods, for salt and bread, for our communities' survival, identity, and dignified existence are struggles that do not take the form of new projects. These are struggles that unfold everyday, at a more or less consistent pace, and in more or less familiar forms. So familiar, in fact, that the analyses of Kamala and Eduardo become organically connected from Khanpur to Oaxaca and give momentum to all of our battles.

GLOSSARY

Daal-roti – Lentils and bread; colloquial for "bread and butter."

Dalit – A member of the scheduled castes, which continue to be subjected to practices of untouchability.

Gur – Jaggery (an unrefined brown sugar) made with cooked sugarcane juice.

Ji – An honorable suffix.

Kadaha – A big iron wok.

Khatolia – A small wooden cot made with woven coir.

Pan-masala – A betel nut–based powder.

Raidas – A Dalit caste specializing in leatherwork.

Roti – Unleavened bread; *chapati.*

Sindur – Vermillion worn in the part of the hair by married Hindu women.

Yatra – Journey.

Sangtin – An Awadhi word denoting solidarity, intimacy, and trust among women companions. With the emergence of Sangtin Kisaan Mazdoor Sangathan in 2005, the term sangtin was embraced by all the members of the movement, irrespective of their gender affiliations.

Sawarn – Upper-caste Hindus who do not fall under the category of "scheduled castes."

NOTES

We thank Sharad Chari, Ananya Chatterjea, Alondra Espejel, Mariano Espinoza, Sylvia Gonzales, Gillian Hart, Rupal Oza, Gerry Pratt, Deborah Rosenstein, Matt Sparke, and Chandra Talpade Mohanty for sowing seeds of some of the above churnings during Surbala's and Richa Singh's visit to the United States in March, April, and May of 2007. We also acknowledge with gratitude the critical feedback from Margalit Chu, David Faust, Sharad Nagar, and Marion Traub-Werner on earlier versions of this chapter in Hindustani and English. Parts of this chapter first appeared in the Hindi weekly *Outlook Saptahik,* in two articles ("Pardesi Zameen par Desi Chhap: Sangtin ki Amerika Yatra thi alag dhang ka Fieldwork," 13 August 2007, 50–51; "Rangbhed aur Naslwad se do-do Haath," 22 October 2007, 46–47).

1. Sangtin Writers, *Playing with Fire: Feminist Thought and Activism Through Seven Lives in India* (Minneapolis: University of Minnesota Press, 2006).

2. The primary sources of support for the movement are the resources provided by SKMS's own membership. In addition, Richa Singh receives a *saathiship* (fellowship for activists) from the Association for India's Development to continue full-time organizing for SKMS. Additional support has come from research funds provided to Richa Nagar by the University of Minnesota to continue her coauthored work with members of SKMS. Some members have occasionally undertaken short-term consultancy work for the government to develop a resource base for SKMS. Royalties from *Sangtin Yatra* and *Playing with Fire,* donations from supporters, and meager profits from a dairy cooperative and a *chikan* embroidery cooperative have also provided some income.

3. The title of the Hindi book *Ek Aur Neemsaar* can be translated as "Another Neemsaar." The literal meaning of "neemsaar" is essence of the *neem,* an ancient tree that is known for its rich medicinal and insecticidal properties and which has been imparted with great

religious, spiritual, and social significance. Despite its bitter taste, the extract of the neem is embraced for all the illnesses that it cures. Neemsaar is also the name of an ancient Hindu religious place of pilgrimage in the Sitapur district. The title, *Ek Aur Neemsaar*, stands for the bittersweet truths of SKMS's journey, a journey that seeks to envision a different world for the people of Sitapur.

4. The touching of feet is an expression of respect. It is important to note that this touching of feet happened during Holi, a day on which it is deemed acceptable to break all taboos. In that sense, it did not carry any "guaranty" of a permanent social shift. However, this event still marked an important breakthrough in our communities.

5. Some authors in the collective chose to use their last names, while others chose to reject them either because they were taking a stand against using last names that reveal caste affiliations or because they did not want their families to feel outed by their own stories and critiques.

6. *Raidas* are seen as lower than Tama's own caste, *Pasi*.

15

TEHRAN KIDS

Mikhal Dekel

Growing up in Haifa, Israel, I knew that my father had survived extermination in Poland by fleeing to British-controlled Palestine during World War II together with a group of children later dubbed the Tehran Children. He was, as I was occasionally told, "lucky" to have been a Tehran Child, but what did that mean? For most of my childhood and much of my adulthood I had not given it much thought. When I did, I had an embarrassingly naive and grossly fantastical picture in my mind of a group of Jewish kids carried en masse from their homes in Poland and brought to safety in Palestine in a kind of instant, relatively painless rescue mission. I knew they were attached to an army—I used to believe it was the Red Army—and that they remained in Tehran for a while. Like most post-Six-Day-War sabras of my generation, I didn't take much interest in the Holocaust and took care to avoid confronting any aspect of it too directly and closely. The Shoah was studied in middle and high school, but in a kind of dry, factual fashion that doesn't leave a mark on one's mind. There was, of course, the annual Holocaust Memorial Day, when we would have school ceremonies and when national television (there was only one channel back then) would broadcast Holocaust films nonstop and radio stations would play somber songs, but I generally did not watch or listen or feel affected by this day. If one looked for it, of course the Holocaust was all over Israeli culture at that time— the works of Appelfeld, Kaniuk, and Kazetnik, not to speak of the Yad Vashem Museum, but these works did not appeal to secular Israeli youth of my generation. There is in Zionist ideology a kind of mandatory rejection of suffering, an imperative to joy above all else (as in the popular post-state song "mukhrakhim lihiyot same'akh"—you *must* be happy), which is really the flipside of extreme sadness. And there has always been in Israeli culture something anarchic, a disinterest in even the nearest past. The Holocaust was abstract, remote and unconnected to our present lives; besides, we had our own (Arab-Israeli) wars— which affected my childhood and adolescence deeply and personally. The dead and wounded of these wars were tragic heroes and not hapless victims, and it was during Memorial Day for the Fallen of the Israeli Defense Forces that I cried my eyes out, glued to those gut-wrenching documentaries on TV about beautiful young men, eighteen-, nineteen- and twenty-year- olds, often talented and unique and as familiar-looking as my own brother.

There was, I think, for people of my generation a dread of the enormity and extreme cruelty of the Holocaust. Many of us were removed by only one generation, which was still not enough to make us feel safe. And my father's generation, who had personally lived through the war—in actuality many of the adults we knew—felt themselves, I think, and rightly so, exempt from watching the representation of the Holocaust in literature and film. It makes sense that if you lived through hell you shouldn't have to watch it on national television, and also, adversely, that the study of hellish mass traumas could be potentially more appealing to scholars with relatively safe lives in America, Europe, and even in present-day Israel.

Nowadays, I also live the safe life of an intellectual far removed from a zone of conflict and mass trauma, so much so that I sometimes wonder if my own relatively recent interest in representations of large-scale catastrophes isn't a product of my comfortable, bloodless, abstract American life. Or is it perhaps the distance and abstraction that my life in America affords me that permits my imagination to get nearer to the source of my father's trauma.

The first thing I discovered after I started doing a bit of research into the story of the Tehran Children was that their journey took three and a half years: from September 1939 to February 1943. *Three and a half years.* One thousand two hundred and seventy seven days of daily battles for a piece of bread, and nearly five hundred of those days spent without parents or other adult protectors. My father, Hannan Dekel (born Tejtel), was thirteen when the Nazis invaded Poland, and his sister Rivka (who back then went by the name Regina) was eight; their cousin Naomi was six. The Tejtels were a provincial aristocracy of sorts, whose successful brewery, I later learned, had supplied beer to America during the Prohibition. They escaped from Poland to Soviet-controlled territories in the days following the Nazi invasion. And in March 1940, six months after they had left, the Soviets presented them with the choice to either rescind their Polish citizenship for a Soviet one or immediately return to Poland. They chose to return.

Instead they were boarded on cattle cars and shipped off in the opposite direction: to a Siberian gulag in the eastern Russian provinces. They traveled for three weeks without food and with little water. In the gulag they were slave laborers, cutting down the trees of the vast Siberian forests in deadly temperatures. At least a quarter of the gulag's inhabitants died there, if not on the way. But after fourteen months, when Germany invaded the USSR, they were released under a general amnesty to Polish citizens. They decided to head south, toward warmer Uzbekistan, walking in the snow for weeks or traveling aboard cargo trains alongside troops of the newly formed Polish army-in-exile (the "Anders Army"). In Tashkent, Uzbekistan, they joined the huge body of naked and starving refugees that were already there.

It was worse than the gulag. For the refugees, Uzbeki towns were full of sickness and violence, and without a shred of food. Because of the heat, typhus and malaria spread quickly and for a while my father's entire family was hospitalized. Plus, they had nothing left to barter with. Scores of kids roamed the streets on their own, including my father's cousin Naomi, whom they accidently found. Nearing starvation, my grandparents decided to place the kids in one of the Christian-Polish orphanages that had been set up across Central Asia. These were not always hospitable to Jewish kids, but they had food; the streets outside their walls were far worse.

Six months later, it was decided that troops of the Anders Army, who were also stationed in Uzbekistan, would be sent to Iran to strengthen British armies in the Middle East. Along with the troops, 11,000 Polish citizens were taken, including 3,000 children who were culled directly out of the orphanages. Among these were the 900 Jewish children who would become known as the Tehran Children, and among them were Hannan, Regina, and Naomi. The Tehran Children ranged from ages one to eighteen. They left Tashkent in trains, sailed for 25 hours on the Caspian Sea and arrived at Bandar Pahlavi, an Iranian port city, in early 1942.

In Iran they were safe for the first time, though still hungry and cold. They lived in Polish army tent camps on the outskirts of Tehran; in the hierarchical ladder that consisted of adult soldiers, adult civilians, Polish children, and Polish-Jewish children, the latter were always at the bottom of the heap. Eventually the Jewish Agency, a Zionist organization (which was operating in Tehran semi-openly) set up a separate "Jewish" camp, where the children remained for seven months.

The Jewish Agency wanted to get the children to Palestine/Eretz Yisrael, but it was nearly impossible, given the British refusal to admit any refugees into Palestine during the war. An initial plan to bus them through Iraq was doomed when Iraq refused at the last minute; another plan to fly them aboard American war planes also fell through. Finally, in January 1943, an agreement was reached with the British and the Egyptians. Overnight and without any preparation, the children were transferred to Bandar Shahpur, where they embarked on a British war ship headed to Karachi and, after a few weeks' stay, on another warship from Karachi to Suez, Egypt. They then crossed the Sinai Desert by train and were quarantined for two days in El Arish; after nearly four years, they arrived in Atlit, a port city located between Haifa and Tel-Aviv. It was February 18, 1943.

My father's parents stayed in Uzbekistan until the end of the war. They survived but remained refugees for years. Even after the end of the war, the British did not permit my grandfather Zindel, who was by then sick with tuberculosis, to join his children in British-controlled Palestine. My grandmother Rachel stayed back with him, probably in Neubeginn, a Displaced Persons' hospital near Munich (and one of the few German DP hospitals to admit TB patients), where he lingered until his death in 1949. That year my grandmother joined her children in the newly founded State of Israel. She had been displaced for ten years, 3,650 days, without even counting her additional struggles in Israel.

This was the outline. I wanted details. Yet in each historical account I read the story was told in an abbreviated form: some months in the Soviet Union, a year in a Siberian gulag, more months in Tashkent, Uzbekistan, close to a year in Tehran. . . . I found the seminal book on the Tehran Children—*Children of Zion*, by a Polish writer named Henryk Grynberg, until then unknown to me. Grynberg had made a collage from testimonies that were collected from the children by Polish officials shortly after their arrival in Palestine. Still, these were abstract, nameless, factual stories. Their amalgamation conveyed the idea that individual narratives are crushed under conditions of extreme brutality, an idea that Primo Levi and other Holocaust memoirists had explicitly articulated. Still, my father was not in the Nazi camps. He was with his family at least some of the time. There were dynamics within the family; they were their distinct selves throughout.

I wanted details but I didn't want to search for them or write an academic book about them. I had this idea that a colleague of mine, the Iranian-born fiction writer Salar Abdoh, would write a novel about the Tehran Children. Salar had become interested in the story after reading about it in a piece by Abbas Milani in the *The Iranian*, a Persian-American journal. I imagined the plot of one main character—a compilation of my father, his sister, and other Tehran children, perhaps—that would stretch over hundreds of pages and counter the thinness and dryness of the testimonies I had read. I believed that Salar, who creates worlds out of his sheer imagination, who is not as mentally blocked by the limitations of scholarly research as I am, and who, most important, is not a direct heir to the grave and weighty legacy of the Holocaust, would do a better job with this material than I would. It's a strange thing for a writer to ask another writer to write her story. But I had a burning desire to understand my father and to have his story told, and yet felt that I could not tell it. I did not want to live with my father for more years of my life. Plus, I felt too close to the burn of his trauma: there is a reason why Amnesty International forbids its volunteers from working on their own countries. And I was also convinced that my father's dramatic story would best be told not by an insider but by an emphatic outsider—Salar—who also of course had the unique advantage of knowing Tehran through and through. Because he came from *there*, from the one place where my father began to feel a bit safe in this world and about which I knew absolutely nothing, I trusted him immediately. And no current Israel-Iran tension or the echoes of the Palestinian–Israeli conflict, which reverberated across and clouded several friendships I had had with other Middle Easterners in New York, could affect that.

I also saw in Salar, whom I soon found out was himself a child refugee of the 1979 Islamic Revolution, something of my father: Microscopic similarities detectable only to one trained by lifelong cohabitation with a refugee: an anxious relationship to food and to cold; the overcautiousness and preparedness of someone who was for a long time exposed to the elements and expects everything to go wrong; but also a very specific humility that emanates from people who have seen too much in their lives, or who have lost everything suddenly, an attribute that stood in utter contrast to many of our self-important colleagues at the university.

I believe other people's stories can open for us a door into our own. My father, for one, was brought to tears not by a Holocaust film but by the post-Vietnam War film *The Deer Hunter*. It was the only time, as far as I can recall, that we saw him crying. He cried at the famous scene in which Nick (Christopher Walken) is played by the Vietcong as the pawn in a Russian roulette game. And though I doubt that my siblings and I could quite grasp the monumental tragic note of that scene, I remember looking at my father—if I recall correctly it was winter and as always slightly chilly in our Haifa apartment—and seeing that his blue eyes were red and tears were rolling down his cheeks.

I don't remember the precise emotion I felt then, but I know it was in the realm of love and pity. My father was a walled-in man, and there was by then already an invisible wall between us, but the display of raw emotion humanized him and made us feel closer to the man. For both my brother and me, independently, the image of my father crying as he watched *The Deer Hunter* remains even today, sixteen years after his death, one of our strongest and most moving memories of him.

Thinking about it so many years later, I realize now that this scene, about an American ex-POW addicted to pain, too brutalized to care that he had become a gambling pawn, spoke to my father deeply and intimately. For a moment it allowed him to release an otherwise bottled-up pain, which usually manifested itself only as anger. And I gradually understood that Salar's story, which is radically different from my father's, began to have for me a similar cathartic function. Increasingly I pondered my father's story through Salar's eyes—grilling him about his own experiences of hunger and the like as a child exile. I also embarked on more serious research, picking up speed as I got deeper into it: gobbling up books, digging up testimonies and archival materials, interviewing former Tehran Children, now in their eighties, about their journeys and subsequent lives in Israel. The harrowing narrative of the Tehran Children, which has hitherto been told and understood in Israel in a predominantly generic and impersonal manner, slowly began to come alive for me in all its vivid sadness and resiliency; meanwhile, for Salar the story of the Tehran Children had become a kind of reference point and window into his own past experiences as a child-refugee of the Islamic Revolution, which had thrust him and his brothers onto the streets of Los Angeles and New York at about the same age that World War II had thrust my father and his sister onto the streets of Tashkent and Tehran.

CONTRIBUTORS

RACHEL ADAMS is professor of English and American studies at Columbia University. She is the author of *Sideshow USA: Freaks and the American Cultural Imagination* (University of Chicago Press, 2001) and *Continental Divides: Remapping the Cultures of North America* (University of Chicago Press, 2009). Her current research is about intellectual disability and the arts.

MIEKE BAL is Royal Netherlands Academy of Arts and Sciences Professor (KNAW). She is based at the Amsterdam School for Cultural Analysis (ASCA), University of Amsterdam. Her areas of interest range from biblical and classical antiquity to seventeenth-century and contemporary art and modern literature, feminism, and migratory culture. Her thirty books include *A Mieke Bal Reader* (University of Chicago Press, 2006), *Travelling Concepts in the Humanities* (University of Toronto Press, 2002), and *Narratology: Introduction to the Theory of Narrative* (University of Toronto Press, 3rd edition 2009). Bal is also a video-artist, and her experimental documentaries on migration include *A Thousand and One Days*; *Colony*; and the installation *Nothing Is Missing*. Occasionally she works as an independent curator.

MIKHAL DEKEL is an associate professor of English and comparative literature at the City College, City University of New York (CUNY). She is the author of *The Universal Jew: Masculinity, Modernity and the Zionist Moment* (Northwestern University Press, 2010). Her articles and translations have appeared in *ELH: English Literary History*, *Women's Studies Quarterly* (*WSQ*), *Callaloo*, and *Guernica*, as well as in various anthologies.

AGNESE FIDECARO is an independent scholar who has taught English, comparative literature, and gender studies at the University of Geneva. She is currently writing a book titled *Exposed Bodies: Crises of Experience in 20th-Century Literature*. Other areas of research include contemporary women's authorship, the politics of language and identity, and women's literary history. She coedited *Le genre de la voix* (*Equinoxe* 23, 2002), *Profession: créatrice: la place des femmes dans le champ artistique* (Antipodes, 2007), and *Femmes écrivains la croisée des langues / Women Writers at the Crossroads of Languages, 1700–2000* (MētisPresses, 2009).

INDERPAL GREWAL is chair and professor of the Women's, Gender and Sexuality Studies Program at Yale University. She is author of *Home and Harem: Nation, Gender, Empire and Cultures of Travel* (Duke, 1996) and *Transnational America: Feminisms, Diasporas, Neoliberalisms* (Duke, 2005); coeditor, with Caren Kaplan, of *Scattered Hegemonies: Postmodernity and Transnational Feminist Practices* (University of Minnesota Press, 1995) and *Introduction to Women's Studies: Gender in a Transnational World* (McGraw Hill, 2001, 2005). Her areas of research include feminist theory, cultural studies of South Asia and its diasporas, and postcolonial and feminist transnationalisms. She is currently working on a book-project titled "Feminism, Security and Postcoloniality in an Age of Terror."

MARIANNE HIRSCH is the William Peterfield Trent Professor of English and Comparative Literature at Columbia University, as well as the codirector of the Center for the Critical Analysis of Social Difference. She is a former editor of *PMLA*. Most recently, she coauthored, with Leo Spitzer, *Ghosts of Home: The Afterlife of Czernowitz in Jewish Memory and History* (University of California Press, 2010). Her other publications include *Family Frames: Photography, Narrative, and Postmemory* (Harvard University Press, 1997) and the forthcoming *The Generation of Postmemory: Visual Cultures After the Holocaust* (Columbia University Press, 2012). She has edited and coedited a number of books, including *The Familial Gaze* (Dartmouth, 1999), *Teaching the Representation of the Holocaust* (Modern Language Association of America, 2005), and *Rites of Return: Diaspora Poetics and the Politics of Memory* (Columbia University Press, 2011).

TSUNG-YI MICHELLE HUANG is an associate professor of geography at National Taiwan University. She is the author of *Walking Between Slums and Skyscrapers: Illusions of Open Space in Hong Kong, Tokyo and Shanghai* (University of Washington Press, 2004). Her current project seeks to explore narratives of female success and spatial imaginations of city connections in contemporary China.

MIN JIN LEE'S debut novel, *Free Food for Millionaires* (Grand Central Publishing, 2007), was a Top 10 Novels of the Year for *The Times* (London), NPR's *Fresh Air*, and *USA Today*. Her short fiction has been featured on NPR's *Selected Shorts*. Her writings have appeared in *Condé Nast Traveler*, *The Times* (London), *Vogue*, *Travel + Leisure*, the *Wall Street Journal*, and *Food & Wine*. Her essays and literary criticism have been anthologized widely. She served as a columnist for the *Chosun Ilbo*, the leading paper of South Korea, for three consecutive terms. She lives in Tokyo with her family and is working on her novel *Pachinko*.

CHI-SHE LI currently teaches at the Department of Foreign Languages and Literatures, National Taiwan University. His published works focus on cultural globalization in late-Victorian England and in contemporary East Asia. Now he is engaged in a project on circuitous connections between mid-Victorian novels and finance capitalism.

NANCY K. MILLER is Distinguished Professor of English and Comparative Literature, the Graduate Center, City University of New York (CUNY). She has published extensively as a feminist literary critic, specializing in women's writing and autobiography. Her books include *Bequest and Betrayal: Memoirs of a Parent's Death* (Indiana University Press, 2000) and *But Enough About Me: Why We Read Other People's Lives* (Columbia University Press, 2002). She is coeditor of *Extremities: Trauma, Testimony, and Community* (University of Illinois Press, 2002), as well as the forthcoming *Rites of Return* and *Picturing Atrocity: Reading Photographs in Crisis*. Her most recent book is a family memoir, *What They Saved: Pieces of a Jewish Past* (University of Nebraska Press, 2011).

GERALDINE PRATT is professor of geography at the University of British Columbia. She is author of *Working Feminism* (Edinburgh University Press, 2004) and *Families Apart: Migrant Mothers and the Conflicts of Labor and Love* (University of Minnesota Press, 2012) and coauthor of *Gender Work and Space* (Routledge, 1995). Her coauthored play, *Nanay: a testimonial play*, was performed in Vancouver and Berlin in 2009.

ELSPETH PROBYN is SA Research Professor of Gender & Cultural Studies, and codirector of the Centre for Postcolonial and Globalization Studies at the University of South Australia. She has written extensively on the lived body, and her books include: *Blush: Faces of Shame* (Minnesota University Press, 2005); *Remote Control: New Media and Ethics*, edited by Catharine Lumby and Elspeth Probyn (Cambridge University Press, 2003); *Carnal Appetites: FoodSexIdentities* (Routledge, 2000);

Outside Belongings (Routledge, 1996); *Sexing the Self: Gendered Positions in Cultural Studies* (Routledge, 1993); and *Creating Value. The Humanities and Public Engagement,* edited by E. Probyn, S. Muecke, and A. Shoemaker (Australian Academy of Humanities, 2006). Her new research focuses on local communities, taste, and place within the transglobal food system and is funded by an Australian Research Council grant.

MARISA BELAUSTEGUIGOITIA RIUS is currently chair of the Gender Studies Program of the National Autonomous University of Mexico (UNAM). She has a doctorate in ethnic studies at the University of California, Berkeley, and a master's degree in psychoanalytic theory and Ibero American literature. Her work focuses on the development of new pedagogies related with precarious spaces such as prisons, transnational feminism, race, gender at the borders of the nation, comparative Latina and Chicana theory, and the relation of art, activism, and academia. She has published several books and articles in national and international journals around those themes. She works currently at the prison of Santa Martha Acatitla in Mexico City in the development of murals as a space for voice, demand, and visibility of the interns and their claim for justice and education. Her main publications are *Enseñanzas deshordadas* (UNAM, 2008) and *Güeras y prietas: género y raza en la construcción de mundos nuevos* (UNAM, 2009).

VICTORIA ROSNER is associate dean of the School of General Studies at Columbia University, where she also teaches in the Department of English. She is the author of *Modernism and the Architecture of Private Life* (Columbia University Press, 2005), awarded the Modernist Studies Association Book Prize. She has guest edited in journals including *WSQ: Women's Studies Quarterly, The Scholar and Feminist Online,* and *Signs: Journal of Women in Culture and Society.* She is currently at work on a book, "Machines for Living: Literature, Modernization, Domesticity," and a memoir.

SIDONIE SMITH is the Martha Guernsey Colby Collegiate Professor of English and Women's Studies at the University of Michigan and the 2010 president of the Modern Language Association of America (MLA). Her fields of specialization include autobiography studies, narrative and human rights, feminist theories, and women's studies in literature. Her most recent books include the second edition of *Reading Autobiography: A Guide to Interpreting Personal Narratives,* coauthored with Julia Watson (University of Minnesota Press, 2001), and *Human Rights and Narrated Lives: The Ethics of Recognition,* coauthored with Kay Schaffer (St. Martin's Press,

2004). She is currently collaborating with four colleagues on a book on autobiographical hoaxes.

ARA WILSON is director of the Program in the Study of Sexualities and associate professor of women's studies at Duke University. She is the author of *The Intimate Economies of Bangkok: Tomboys, Tycoons and Avon Ladies in the Global City* (University of California Press, 2004) and is currently completing *Sexual Latitudes: The Erotic Life of Globalization*, an interdisciplinary exploration of globalization as a condition for sexuality. Her new research on the political economy of intimacy looks at medical tourism to Thailand and has been published in Ong and Chen, eds., *Asian Biotech* (University of California Press, 2010) and *Body and Society* (2011).

MELISSA W. WRIGHT teaches in the Department of Geography and the Department of Women's Studies at the Pennsylvania State University. Her publications include *Disposable Women and Other Myths of Global Capitalism* (Routledge, 2006) and a number of articles in geography, cultural studies, and feminist journals. She has conducted research on social activism, capitalist exploitation, and urban politics in northern Mexico for twenty years and is now investigating the experience of violence and militarization on border communities along both sides of the Mexico-U.S. border.

SANGTIN WRITERS

REENA'S activism began in 1998 as an instructor in a women's literacy center in the Mahila Samakhya Programme of Sitapur District in the Indian state of Uttar Pradesh. She became a board member of Sangtin in 2003 and emerged as an effective mobilizer in the villages. Reena founded a dairy cooperative in the Kunwarapur village and worked to integrate it with the movement-building activities of Sangtin Kisaan Mazdoor Sangathan (Sangtin Peasants' and Laborers' Organization). Reena works on women's political participation, social violence, and rural livelihoods and has traveled in several states of India to participate in dialogues on these subjects.

RICHA NAGAR is professor of gender, women and sexuality studies at the University of Minnesota and a founding member of Sangtin Kisaan Mazdoor Sangathan in Sitapur, India. She coauthored *Sangtin Yatra* (2004), *Playing with Fire: Feminist Thought and Activism Through Seven Lives in India* (University of Minnesota Press, 2006), and *A World of Difference: Encountering and Contesting*

Development (Guilford Press, 2009), and coedited *Critical Transnational Feminist Praxis* (SUNY Press, 2010). Richa's research on gender, race, and communal politics among South Asian communities in postcolonial Tanzania and her subsequent work have resulted in numerous articles. Since 1996, her research, organizing, and creative work have focused mainly on collaborations that seek to reconfigure the political terrain and processes associated with "empowerment" projects aimed at "the poor."

RICHA SINGH, a founding member of Sangtin Kisaan Mazdoor Sangathan (SKMS) in Sitapur, India, stepped into activism in 1991 as a member of the office staff in Mahila Samakhya (MS) in Banaras and later emerged as a mobilizer and district coordinator in the Saharanpur and Sitapur programs of MS. In MS, Richa participated in dialogues on gendered violence and informal education on regional and national platforms. The coauthorship of *Sangtin Yatra* marked the beginning of her full-time movement-building work with peasants and workers of Sitapur to build SKMS, which currently focuses on access to irrigation and livelihoods, and on socioeconomic violence.

SURBALA, a coauthor of *Sangtin Yatra,* is a founding member of Sangtin, now Sangtin Kisaan Mazdoor Sangathan (SKMS). Since 1996, her activism has focused on mobilizing poor women, peasants, and laborers in the villages of India's Sitapur District. After working in Mahila Samakhya-Sitapur for five years, Surbala left her job to begin full-time organizing for Sangtin. She formed a *chikan* embroiderers' collective and worked on issues ranging from gendered violence, casteism, and communalism to rural people's livelihoods. Surbala has traveled in several states of India and the United States to participate in critical dialogues on these subjects.

INDEX

GENDER AND CULTURE

A SERIES OF COLUMBIA UNIVERSITY PRESS

Nancy K. Miller and Victoria Rosner, Series Editors

Carolyn G. Heilbrun (1926–2003) and Nancy K. Miller, Founding Editors

In Dora's Case: Freud, Hysteria, Feminism
EDITED BY CHARLES BERNHEIMER AND CLAIRE KAHANE

Breaking the Chain: Women, Theory, and French Realist Fiction
NAOMI SCHOR

Between Men: English Literature and Male Homosocial Desire
EVE KOSOFSKY SEDGWICK

Romantic Imprisonment: Women and Other Glorified Outcasts
NINA AUERBACH

The Poetics of Gender
EDITED BY NANCY K. MILLER

Reading Woman: Essays in Feminist Criticism
MARY JACOBUS

Honey-Mad Women: Emancipatory Strategies in Women's Writing
PATRICIA YAEGER

Subject to Change: Reading Feminist Writing
NANCY K. MILLER

Thinking Through the Body
JANE GALLOP

Gender and the Politics of History
JOAN WALLACH SCOTT

The Dialogic and Difference: "An/Other Woman" in Virginia Woolf and Christa Wolf
ANNE HERRMANN

Plotting Women: Gender and Representation in Mexico
JEAN FRANCO

Inspiriting Influences: Tradition, Revision, and Afro-American Women's Novels
MICHAEL AWKWARD

Hamlet's Mother and Other Women
CAROLYN G. HEILBRUN

Rape and Representation
EDITED BY LYNN A. HIGGINS AND BRENDA R. SILVER

Shifting Scenes: Interviews on Women, Writing, and Politics in Post-68 France
EDITED BY ALICE A. JARDINE AND ANNE M. MENKE

Tender Geographies: Women and the Origins of the Novel in France
JOAN DEJEAN

Unbecoming Women: British Women Writers and the Novel of Development
SUSAN FRAIMAN

The Apparitional Lesbian: Female Homosexuality and Modern Culture
TERRY CASTLE

George Sand and Idealism
NAOMI SCHOR

Becoming a Heroine: Reading About Women in Novels
RACHEL M. BROWNSTEIN

Nomadic Subjects: Embodiment and Sexual Difference in Contemporary Feminist Theory
ROSI BRAIDOTTI

Engaging with Irigaray: Feminist Philosophy and Modern European Thought
EDITED BY CAROLYN BURKE, NAOMI SCHOR, AND MARGARET WHITFORD

GENDER AND CULTURE READERS